CLASSICAL APPROACH

to

CONSTRAINED AND UNCONSTRAINED
MOLECULAR DYNAMICS

AJITH GUNARATNE

To order additional copies of this book, contact:
Xlibris
1-888-795-4274
www.Xlibris.com
Orders@Xlibris.com

DEDICATION

This work is dedicated to my loving wife Nilanthi Gunaratne, energetic, bright, boundless sons Chamara Gunaratne and Chanith Gunaratne, who has been a constant source of support and encouragement and trust, I would not have been able to complete this work. . I am truly thankful for having you in my life. This work is also dedicated to my parents, Mr and Mrs Gunaratne, who have always loved me unconditionally and whose good examples have taught me to work hard for the things that I aspire to achieve.

CONTENTS

LIST OF TABLES   4

LIST OF FIGURES   5

ABSTRACT   9

CHAPTER 1.
INTRODUCTION AND BACKGROUND   11

1.1 Introduction   11

1.2 Background   13

    1.2.1 Protein   16

1.3 Empirical force field   16

    1.3.1 Introduction   17

    1.3.2 Bond stretching potential - φb 18

    1.3.3 Angle bending potential - $\varphi \theta$ 19

    1.3.4 Torsion potential - $\varphi \tau$ 20

    1.3.5 Potential of non-bonding interactions - φnb 21

1.4 Molecular dynamic simulations   23

    1.4.1 Introduction   23

    1.4.2 History   24

    1.4.3 Limitations   25

1.5 Unconstrained molecular dynamic simulations   27

    1.5.1 Verlet algorithm   27

    1.5.2 Leap-Frog algorithm   30

    1.5.3 Predictor-Corrector algorithm   31

    1.5.4 Velocity version of Verlet algorithm   32

    1.5.5 Beemans algorithm   33

    1.5.6 Symplectic integrators   33

1.6 Constrained molecular dynamic simulations   34

    1.6.1 Shake algorithm   34

1.6.2 Rattle algorithm   35

1.6.3 Stochastic method   38

1.6.4 Velocity rescaling   38

1.7 Review   39

CHAPTER 2.
LAGRANGE MULTIPLIER METHOD   40

2.0.1 Lagrange multiplier method   40

2.0.2 Time dependent Lagrange multiplier method for molecular dynamics   41

2.1 Review   45

CHAPTER 3.
PENALTY AND BARRIER METHODS   46

3.0.1 History   46

3.0.2 Constraints   47

3.0.3 Penalty function method   48

3.0.4 Karush-Kuhn-Tucker multipliers   52

3.0.5 Exact penalty function   54

3.0.6 Barrier method   56

3.1 Review   59

CHAPTER 4.
MOLECULAR DYNAMICS, PENALTY FUNCTION METHOD AND ITS PROPERTIES   60

4.0.1 Constrained molecular dynamics and penalty function method   60

4.1 Analysis of molecular dynamics   70

4.1.1 Root Mean Square Deviation (RMSD)   70

4.1.2 Velocity Autocorrelation Function (VAF)   72

4.1.3 Ramachandran Plots   74

4.2 Review   75

CHAPTER 5.
IMPLEMENTATION PROCEDURE   76

5.1 Introduction   76

5.1.1 Penalty function implementation on Argon clusters   77

5.2 CHARMM settings   84

    5.2.1 CHARMM minimization energy process   89

    5.2.2 Minimization methods   91

    5.2.3 CHARMM force field   93

    5.2.4 Convergence criteria   94

5.3 Penalty method implementation   95

5.4 Review   101

CHAPTER 6.
RESULTS, SUMMARY AND DISCUSSION   102

    6.0.1 Analysis of dynamics   102

6.1 Review   107

CHAPTER 7.
EVALUATION/CONCLUSION   108

APPENDIX A.
FORTRAN PROGRAM FOR PENALTY FUNCTION METHOD FOR ARGON CLUSTERS   110

A.0.1 Main program   110

A.0.2 Sub program Verlet   114

A.0.3 Sub program Init velocity   116

A.0.4 Sub program read files   118

A.0.5 Sub program distance   122

APPENDIX B.
HIGH PERFORMANCE FORTRAN PROGRAM FOR PENALTY FUNCTION METHOD FOR ARGON CLUSTERS   123

B.0.1 Sub program - Verlet   129

B.0.2 Sub program - Position Init   131

B.0.3 Sub program - Init velocity   135

B.0.4 Sub program - bubble sort   137

BIBLIOGRAPHY   138

LIST OF TABLES

Table 5.1 Final steps of energy minimization   98

Table 5.2 Computing time of VL, SH and PL run. *Computing time for the 25ps simulation after equilibrium 100

Table 5.3 Root mean square deviation (RMSD) of backbone atoms   101

LIST OF FIGURES

Figures 1.1   The chemical formulas of 20 amino acids (47). Plot is created by Chemsketch software. (Advanced Chemistry Development Lab - www.acdlabs.com).   14

Figures 1.2   The space filling model of 20 amino acids. VMD visualization software is used. Color is based on ResID.   15

Figures 1.3   Three dimensional structure of Bovine Pancreatic Trypsin Inhibitor (BPTI) protein with 58 residuals. Data are downloaded from protein data bank (PDB) which released on 18-Jan-1983 (7). VMD visualization software is used. Color is based on ResID.   17

Figures 1.4   Bond stretching potential energy.   18

Figures 1.5   Angle bending potential energy.   20

Figures 1.6   Torsion potential energy.   21

Figures 1.7   Lennard Johnes potential of single pair of atoms.   22

Figures 5.1   The figure is illustrated potential energy changes when penalty term change.   79

Figures 5.2   The flow chart of the penalty function algorithm for Argon cluster simulation.   80

Figures 5.3   Changes in potential energy of the trajectory for argon cluster 13 produced by the penalty function method. Here, randomly selected 60% of all distances were constrained to their distances in the global energy minimum configuration. The trajectory already approached to the global energy minimum (-44.3) of the cluster in 3000 time steps while the trajectory generated by the Verlet remained in high energy. The time step is $0.032ps$ and penalty term updated every 500 iteration by 1.   81

Figures 5.4   Changes in potential energy of the trajectory for argon cluster 13 produced by the penalty function method. Randomly selected 60% of all distances were constrained to their distances in the global energy minimum configuration. The time step is $0.032ps$ and penalty term updated every 1000 iteration by 5.   82

Figures 5.5   Changes in potential energy of argon cluster 24. Solid and dotted lines show the potential energy of the trajectory produced by the Verlet (VL) and penalty function (PL) methods, respectively. Here, randomly selected 50% of all distances were constrained to their distances in the global energy minimum configuration (-97.349).   83

Figures 5.6   CHARMM simulation procedure.   85

Figures 5.7   Basic steps of molecular dynamic simulation procedure.   88

Figures 5.8   Initial BPTI structure downloaded from PDB data bank. Picture uses display style cartoon, coloring is based on RESID and use VMD software.   89

Figures 5.9   BPTI with four water molecules. Picture uses display style CPK. VMD software is used to create picture. Color is based on RESID.   90

Figures 5.10   The figure is showed sequence of BPTI (9).   95

Figures 5.11   This picture shows BPTI with all hydrogen atoms. There are 904 atoms in total. Picture uses display style CPK and coloring is based on RESID.   96

Figures 5.12   This picture shows minimized BPTI stricture with all hydrogen atoms. Display style is CPK and coloring is based on RESID. VMD is used.   98

Figures 5.13   Average of $25ps$ structure of equilibrium period of BPTI structure including all hydrogen atoms. CPK display style and color is based on RESID. The picture is created by using VMD software.   99

Figures 5.14   Simulation time for VL, SH and PL. *Heating - bring the system to normal temperature; §Equilibrium - the time for the system to reach the equilibrium; ¶Production - stable dynamic results for analysis.   100

Figures 6.1   Temperature distribution of Shake and Penalty run.   102

Figures 6.2   Temperature distribution of Verlet run.   103

Figures 6.3   The average backbone RMS fluctuations of the residues in the $25ps$ production simulations.   103

Figures 6.4   The average C_{α} RMS fluctuations in the $25ps$ production simulations.   104

Figures 6.5   The average RMS fluctuations of the HN atoms in the $25ps$ production simulations.   105

Figures 6.6   The average RMS fluctuations of the non-backbone atoms in the $25ps$ production simulations. 105

Figures 6.7   The velocity auto correlations of the C_α atom of 51 CYS based on the trajectories produced by VL, SH, and PL in a time period of $0.1ps$.   106

ABSTRACT

We propose a penalty-function method for constrained molecular dynamic simulation by defining a quadratic penalty function for the constraints. The simulation with such a method can be done by using a conventional, unconstrained solver only with the penalty parameter increased in an appropriate manner as the simulation proceeds. More specifically, we scale the constraints with their force constants when forming the penalty terms. The resulting force function can then be viewed as a smooth continuation of the original force field as the penalty parameter increases. The penalty function method is easy to implement and costs less than a Lagrange multiplier method, which requires the solution of a nonlinear system of equations in every time step. We have first implemented a penalty function method in CHARMM and applied it to protein Bovine Pancreatic Trypsin Inhibitor (BPTI).We compared the simulation results with Verlet and Shake, and found that the penalty function method had high correlations with Shake and outperformed Verlet. In particular, the RMSD uctuations of backbone and non-backbone atoms and the velocity auto correlations of $C\alpha$ atoms of the protein calculated by the penalty function method agreed well with those by Shake. We have also tested the method on a group of argon clusters constrained with a set of inter-atomic distances in their global energy minimum states. The results showed that the method was able to impose the constraints effectively and the clusters tended to converge to their energy minima more rapidly than not confined by the constraints.

CHAPTER 1.
INTRODUCTION AND BACKGROUND

1.1 Introduction

Molecular dynamics simulation can be used to study many different dynamic properties of proteins, but a long sequence of iterations has to be carried out even for small protein motions due to the small time step (1.0^{-15} Sec) required (47). The bonding forces are among those causing fast protein vibrations that require small time steps to integrate, but they may be replaced by a set of bond length constraints, to increase the step size and hence the simulation speed (23). Several Lagrange multiplier types of methods have been developed for constrained molecular dynamics simulation. However, in all these methods, the multipliers have to be determined in every time step by solving a nonlinear system of equations so that the new iterate can satisfy the constraints (3). Depending on the number of constraints, the additional computational cost can be large, given the fact that the force field calculation in every time step is at most $O(n^2)$, while the solution of the nonlinear system of equations may require $O(m^3)$, where n is the number of particles in the system and m the number of constraints. In this book, we propose a so-called penalty function (34) method for constrained molecular dynamics. In this method, a special function is defined so that the function is minimized if the constraints are satisfied. By adding such a function in the potential energy function, the constraints can then be removed from the system, and the simulation can be carried out in a conventional, unconstrained manner. The advantage of using a penalty function method is that it is easy to implement, and does not require solving a nonlinear system of equations in every time step. The disadvantage of the method is that the penalty parameter, i.e., the parameter used to scale the penalty function, is hard to control and in principle, needs to be large enough for the penalty function to be truly effective, which on the other hand, may cause numerical instabilities when used in simulation (16). It may also arguably be a disadvantage that the penalty function method only forces the constraints to be satisfied approximately but not completely. The method could be used as an alternatively and computationally more efficient approach for constrained molecular dynamics simulation than the Lagrange multiplier types of methods. We have first implemented a penalty function method in CHARMM (9) and tested it on protein Bovine Pancreatic Trypsin Inhibitor (BPTI) by following a similar experiment done by Gunsteren and Karplus in (23) for the Shake algorithm (38). In this implementation, we removed the bond length potentials from the potential energy function and introduced the corresponding bond length constraints. For each of the bond length constraints, we constructed a quadratic penalty function and inserted it into the potential energy function. For each different type of bond, we also scaled the corresponding penalty function with the force constant of the bond so that the resulting function had the same form as the original bond length potential if without multiplied by the penalty parameter. The resulting force field becomes simply a continuation of the original force field as the penalty parameter changes continuously from 1 to a value > 1. We conducted a simulation on BPTI with the penalty function method, and compared the results with Verlet and Shake, and found that the penalty function method had a high correlation with the Shake and outperformed the Verlet. In particular, the root-mean-square-deviations (RMSD) of the backbone and non-backbone atoms and the velocity auto correlations of the $C\alpha$ atoms of the protein calculated by the penalty function method agreed well with those by Shake. Note again that the penalty function method requires no more than just applying a conventional, unconstrained simulation algorithm such as the Verlet algorithm to the potential energy function expanded with additional penalty terms for the bond length constraints. We have also tested the penalty function method on a group of argon clusters with the equilibrium distances for a selected set of molecular pairs as the constraints. The equilibrium distances mean that distances for the pairs of argon molecules when the clusters are in their global energy minimal states. We generated these distances by using the global energy minimal configuration of the clusters published in previous studies (36). A penalty function was constructed for each of the constraints and incorporated into the potential energy function of the cluster. The simulation was then conducted by using a conventional, unconstrained simulation method, i.e., the Verlet algorithm (49), with the extended potential energy function. There were no substantial algorithmic changes or computational overheads required due to the addition of the constraints. The simulation results showed that the penalty function method was able to impose the constraints effectively and the clusters tended to converge to their lowest energy equilibrium states more rapidly than not confined by the constraints.

We introduce protein, empirical force field, history of molecular dynamics, unconstrained and constrained dynamics in chapter 1. In chapter 2, we present time independent and dependent Lagrange multipliers. Theory of penalty and barrier methods are described in chapter 3 (as a optimization problem). We introduce theory of penalty function methods and statistical properties in chapter 4. Then, in chapter 5, we present Argon simulation and summery of CHARMM program basics followed by penalty function implementation on CHARMM. In chapter 6, we present the results on BPTI and their comparisons with the Verlet and the Shake. We conclude the research in chapter 7. Serial and parallel code of algorithm is presented in appendix A and B.

1.2 Background

One of the simplest ways to describe problems in computational chemistry, yet most difficult to solve is the determination of molecular conformation. A molecular conformation problem can be described as finding the global minimum of a suitable potential energy function, which depends on relative atom positions. Progress toward solution techniques will facilitate drug design, synthesis and utilization of pharmaceutical and material products. The success of computational methods to solve such kind of problems hinges on two factors: a suitable potential energy function to predict the native states of the system as the global minimizer of the potential energy function and the available minimization algorithms that can be used to locate efficiently the global minimizer of the potential energy function. The methods of quantum chemistry are quite suited to predict the geometric, electronic and energy features of known and unknown molecules. However, it remains too expensive in terms of computer time and nearly intractable, even at the simplest, semi-empirical level, for many organic molecules or biological macromolecular structures. Therefore, increased interest has focused on models that are able to give quickly an energy favorable conformation for large systems.

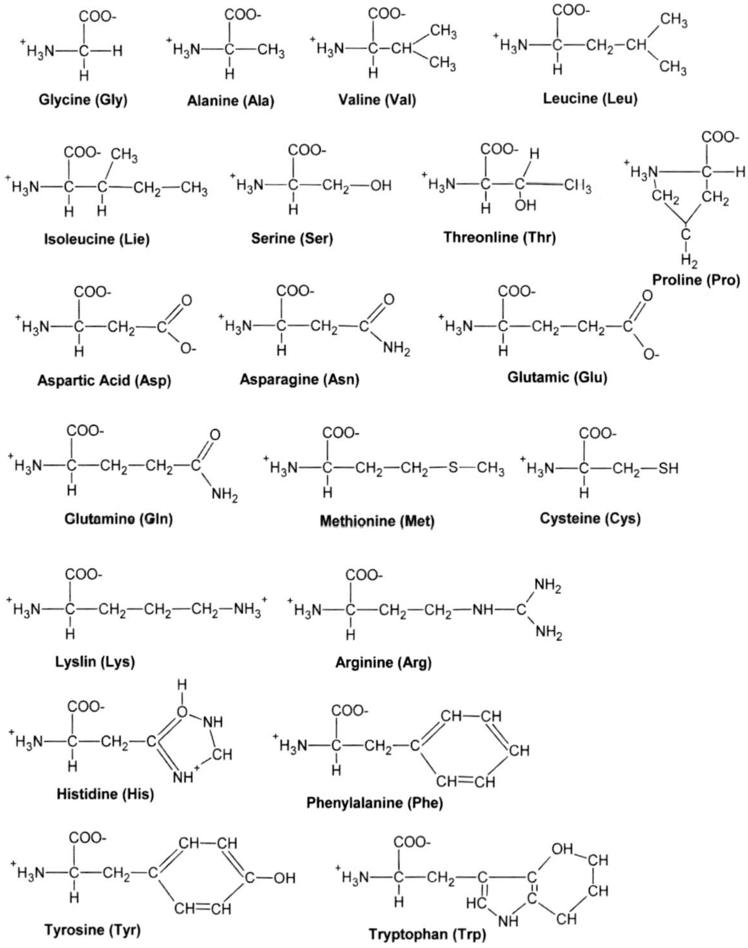

Figure 1.1 The chemical formulas of 20 amino acids (47). Plot is created by Chemsketch software. (Advanced Chemistry Development Lab - www.acdlabs.com).

Molecular mechanics or empirical force field methods are techniques that play an important role in the research of molecular conformation (47).

In a molecular dynamics simulation, the classical equations of motion for the positions, velocities, and accelerations of all the atoms and molecules are integrated forward in time using finite-difference algorithms. The dynamical trajectories given by Newton's equations of motion are approximately calculated (43).

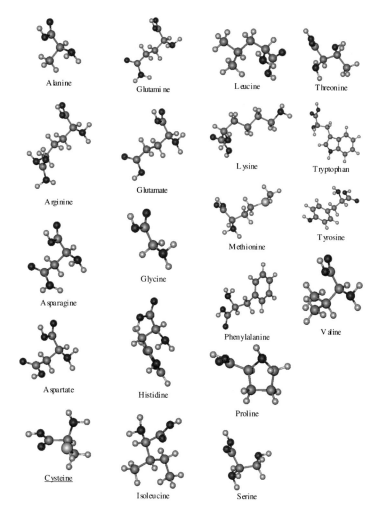

Figure 1.2 The space filling model of 20 amino acids. VMD visualization software is used. Color is based on ResID.

In simulations, we assume that the forces on particles are nearly constant over very short periods of times (*femtosecond* = 10^{-15} *seconds*). During that time, we move the particles along simple parabolic trajectories while recalculating the forces. Then, repeat this process. Most experimental work is done under conditions of constant temperature, constant volume or constant pressure. The main strengths of molecular dynamics are that they efficiently sample the given ensemble, and that they provide dynamical quantities, such as velocity autocorrelation functions, dynamic scattering factors, and diffusion constants. The main weakness of molecular dynamics is an inability to access very long time scales, on the order of one microsecond (10^{-6} seconds) or greater (31).

1.2.1 Protein

Proteins (figure 1.3) are large, complex molecules made from different amino (figure (1.1), (1.2)) acids bonded together sequentially such that they form a long string of a molecule. And like a string, these long molecules can twist and turn and bunch up to have a final shape that is round. These strings actually fold up into distinct structures that usually end up looking overall like a globular structure which are very complex. There are 20 (figure (1.1), (1.2)) amino acids, 9 have sidechains capable of forming hydrogen bonds with each other. There are 2 amino acids with sidechains that can form covalent bonds with each other. The remaining 9 amino acids are water-fearing, and cannot form any kind of bond with each other, but their desire to be away from the external environment of water is a strong force that pushes them towards the inside of the protein [(12), (13), (14)].

1.3 Empirical force field

Empirical forces are played major part of the classical molecular dynamics. The accurate force field is very important for accuracy of the dynamics. First empirical force field functions are discussed in details. Then history of molecular dynamic simulations are presented followed by unconstraint and constraints methods.

1.3.1 Introduction

The goal of molecular modelling is to predict the energy associated with a given conformation of a molecule. The energy of a target molecule depends on the relative positions of its atoms (29). This energy can be approximately estimated by the sum of several contributions.

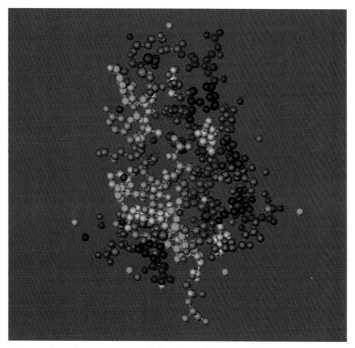

Figure 1.3 Three dimensional structure of Bovine Pancreatic Trypsin Inhibitor (BPTI) protein with 58 residuals. Data are downloaded from protein data bank (PDB) which released on 18-Jan-1983 (7). VMD visualization software is used. Color is based on ResID.

The deformation (23) due to interaction between two non-bonded atoms represents the action of Van der Waals attraction, steric repulsion and electrostatic attraction-repulsion on these two atoms the potential energy function can be studied as a sum of different type of potential term that can be written as (28):

$$\varphi \;=\; \varphi b + \varphi \theta \;+\; \varphi \tau + \varphi nb \;+\; \text{(specific terms)} \tag{1.1}$$

where φ is often referred to as the steric energy or potential energy. It corresponds to the energy difference between the real molecule and a hypothetical molecule in which all structural values, such as bond lengths and bond angles are exactly equilibrium values. In equation (1.1):

- φb {bond energy, describing the compression or the extension of a bond from its equilibrium length.

- $\varphi \theta$ {angle bending energy, and is the function of bond curve with respect to its equilibrium value.

- $\varphi \tau$ {torsion energy.

- φnb {interaction energy between two non-bonded atoms.

- specific terms {could be out of plane bending, electrostatic interactions and possible hydrogen bonding, mean force potential.

1.3.2 Bond stretching potential - φb

The bond stretching contribution (figure 1.4) is represented by Hookes law. It measures the energy due to the variation of bond length after extension or compression from their equilibrium values [(28), (23)]:

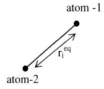

Figure 1.4 Bond stretching potential energy.

$$\varphi b = \frac{1}{2} \sum_{i=1}^{l} k_i \left[r_i - r_i^{eq} \right]^2 \qquad (1.2)$$

where

l - total number of bonds in the molecule

k_i - bond force constant

r_i - bond length

r_i^{eq} - is the bond length at equilibrium position

The parameters k_i and r_i^{eq} are invariant, depending only on the type of each pair of connected atoms. Equation (1.2) is a rough approximation of bond energy. Alternatively, a Morse potential can be used to describe more precisely (29) the bond stretching energy due to the variation of a bond length:

$$\varphi b = \sum_{i=1}^{l} D \left(1 - e^{-a\left[r_i - r_i^{eq} \right]} \right)^2 \tag{1.3}$$

where D and a are parameters characterizing the bond. The use of such a potential seems to be useful for elongated hydrogen bonds, which otherwise tend to dissociate.

1.3.3 Angle bending potential - $\varphi\theta$

Angle bending potential (figure 1.5) determines the energy quantity resulted by [(29), (23)] the angle variation between two adjacent bonds based on an equilibrium bond angle. In the case of harmonic approximation, this is equally derived from Hooks law:

$$\varphi\theta = \frac{1}{2} \sum_{i,j=1}^{n} k_{i,j} \left[\theta_{eq} - \theta_{i,j}^{eq} \right]^2 \tag{1.4}$$

where

$k_{i,j}$ - force constant

$\theta_{i,j}$ - bond angle between 3 atoms

$\theta_{i,j}^{eq}$ - bond angle at equilibrium position between 3 atoms

n - is number of atoms

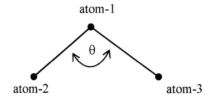

Figure 1.5 Angle bending potential energy.

1.3.4 Torsion potential - $\varphi\tau$

Torsion energy (figure 1.6) represents the energy modification of the rotation of the molecule around a bond. The most common expression which permits to (28) describe the evaluation of the molecule energy as the function of torsion angle is the Fourier series (9):

$$\varphi\tau = \frac{1}{2}\sum_{i=1}^{n} A_{i,s}\left[1+\cos\left(s\tau_i - \Phi\right)\right] \tag{1.5}$$

where

$A_{i,s}$ - force constant which controls the curve amplitude

τ_i - torsion angle

Φ - phase

s - periodicity of $A_{i,s}$

Torsion energy is in fact a correction of different energy terms rather than a physical process. It represents the energy quantity that should be added to or subtracted from the summation of $\varphi b + \varphi \theta + \varphi_{nb}$. Torsion energy is used to obtain the (9) geometry in good agreement with an experiment or with the geometry that is deduced from quantum chemical calculations.

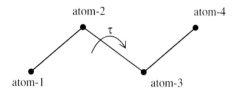

Figure 1.6 Torsion potential energy.

1.3.5 Potential of non-bonding interactions - φnb

Interaction between two non-bonding atoms is the principal cause of steric hindrance, which play an important role in the molecular geometry. The energy of non-bonding interactions is the sum of energies of all non-bonding atoms acting between them (9). It includes the energy of Van der Waals interaction, electrostatic energy and induction energy terms. The term Van der Waals interaction is generally described by the Lennard Johnes potential (figure 1.7):

$$\varphi vdw = \sum_{i<j}^{n} \left[\frac{A_{i,j}}{r_{i,j}^{12}} - \frac{B_{i,j}}{r_{i,j}^{6}} \right] \tag{1.6}$$

where

$A_{i,j}$ and $B_{i,j}$ - are Van der Waals constants

$r_{i,j}$ - is distance between two non-bonding atoms i and j

The summation is taken over all non-bonded pairs of atoms (i, j). These expressions involve two terms:

1. An attractive part, corresponding to induced dipole-induced dipole interaction, proportional to $r_{i,j}^6$, where $r_{i,j}$ is the distance between the two atoms i and j.

2. A repulsive part $r_{i,j}^{12}$, rapidly growing as the distance is getting shorter.

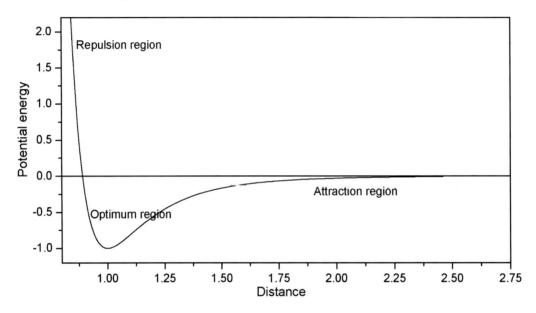

Figure 1.7 Lennard Johnes potential of single pair of atoms.

For a given geometrical arrangement of the atoms in a molecule system, the steric energy, due to distortions of bond lengths and angles with respect to the reference values and Van der Waals interaction (9), can be calculated according to the potential energy function. To determine the actual equilibrium geometry, this steric energy with respect to all internal degrees of freedom must be minimized.

Electrostatic energy increases with the polarity of chemical bonds. It can be expressed using Coulomb potential. Induction energy is the consequence of the distortion of electronic distribution, which depends on the electric field created by other molecules, and generates induced electric moments.

Bond lengths and bond angles are usually available from existing structural information (i.e., from X-ray crystallography). Bond stretching parameters can be directly derived from vibrational force constants. The coefficients of the torsion barriers can be estimated from barrier heights obtained through microwave spectroscopy, thermodynamic studies, or far infrared and Raman spectroscopy. More challenging is the evaluation of the Van der Waals interaction, a crucial point since these interactions are important in determining the stability of crowded or highly branched molecules such as peptides [(9), (29), (28), (47)].

1.4 Molecular dynamic simulations

1.4.1 Introduction

Molecular dynamics has been used for decades to investigate dynamical properties of molecules, solids, and liquids by numerical simulations. In the classical (or conventional) molecular dynamics approach, a model of interatomic interactions must be provided as input before a simulation can be carried out. Such models, or interatomic potentials, are based on a previous knowledge of the physical system studied. Ionic forces can be derived from such model potentials, and atomic trajectories are computed by integrating the Newtonian equations of motion (33).

Due to the vast improvements in computer power, speed, and availability over the past decades the Molecular Dynamics methods are becoming increasingly common technique of simulating molecular scale models of matter. It is now reasonable and possible to simulate realistic (37), large scale blocks of atoms (21) and observe macroscopic (20) effects from these simulations using a desktop computer. In simple terms, a molecular dynamics simulation amounts to finding a numerical solution to the n-body problem. Given a function describing the potential energy the equations of motion can be iteratively solved in order to dynamically simulate the motions of the particles within the system. Next, we save average values for physical, thermodynamic over long time periods. Higher order numerical approximations have always been available. However, they have frequently been passed over in favor of lower order techniques in order to save on computing time. With the massive increases in computational power becoming readily available in smaller and smaller machines one must begin to reevaluate these decisions and begin to bring higher numerical accuracy back into the picture. Whereas before, in order to simulate realistically sized blocks of atoms it was necessary to use a second or third order accurate method (18).

In molecular dynamics, we follow the laws of classical mechanics, and most notably Newton's law (47):

$$m_i a_i = f_i$$

(1.7)

for each atom *i* in a system constituted by *n* atoms. Here, m_i is the mass of the atom, $a_i = \frac{d^2 x_i}{dt^2}$ its acceleration, and f_i the force acting upon it, due to the interactions with other atoms and $x_i = (x_{1i}; x_{2i}; x_{3i}) \in \mathbb{R}^3$. This concludes that molecular dynamics is a deterministic technique. For example, given an initial set of positions and velocities, the subsequent time evaluation is completely determined.

1.4.2 History

There are some of the key papers that appeared in the 50s and 60s which can be regarded as milestones in molecular dynamics. The first paper reporting a molecular dynamics simulation was written by (1). The purpose of the paper was to investigate the phase diagram of a hard sphere system, and in particular the solid and liquid regions. In a hard sphere system, particles interact via instantaneous collisions, and travel as free particles between collisions. The calculations were performed on a UNIVAC and on an IBM 704. The (19) are probably the first example of a molecular dynamics calculation with a continuous potential based on a finite difference time integration method. The calculation for a 500-atoms system was performed on an IBM 704, and spent about a minute per time step. Aneesur Rahman at Argonne National Laboratory has been a well known pioneer of molecular dynamics. In his paper (39), he studies a number of properties of liquid Argon, using the Lennard-Jones potential on a system containing 864 atoms on a CDC 3600 computer. The legacy of Rahman's computer codes is still carried by many molecular dynamics programs in operation around the world.

Loup Verlet calculated (49) the phase diagram of argon using the Lennard-Jones potential, and computed correlation functions to test theories of the liquid state. The bookkeeping device which became known as Verlet neighbor list was introduced in his paper. This method is still popular in unconstrained molecular dynamics. This schema is called Verlet algorithm. Phase transitions in the same system were investigated by Hansen and Verlet in 1969 (25). The velocity version of Verlet is introduced in 1982 (46). Later constraints algorithms are introduced. Shake and Rattle algorithms are widely used constrained algorithms in literature.

1.4.3 Limitations

Molecular dynamics is a powerful technique but has limitations. One weakness is the complication of how we can use Newton's law to move atoms, when the systems at the atomistic level obey, quantum laws rather than classical laws. It has been shown (24) that the classical approximation is poor for very light systems such as H_2, H_e and N_e.

In molecular dynamics, atoms interact with each other. These interactions originate forces which act upon atoms, and atoms move under the action of these instantaneous forces. As the atoms move, their relative positions change and forces change as well. The essential ingredient containing the physics is therefore constituted by the forces. A simulation is realistic only to the extent that interatomic forces are similar to those that real atoms would experience when arranged in the same configuration. Forces are usually obtained as the gradient of a potential energy function, depending on the positions of the particles. The realism of the simulation therefore depends on the ability of the potential chosen to reproduce the behavior of the material under the conditions at which the simulation is run (50).

Typical molecular dynamic simulations can be performed on systems containing thousands or, perhaps, millions of atoms, and for simulation times ranging from a few picoseconds (10^{-12} seconds) to hundreds of nanoseconds (10^{-9} seconds). While these numbers are certainly respectable, it may happen to run into conditions where time and/or size limitations become important.

The engine of a molecular dynamics program is its time integration algorithm, required to integrate the equation of motion of the interacting atoms and follow their trajectory. Time integration algorithms are based on finite difference methods, where time is discretized on a finite grid, the time step Δt being the distance between consecutive points on the grid. Knowing the positions and some of their time derivatives at time t, the integration scheme gives the same quantities at a later time $(t+\Delta t)$. By iterating the procedure, the time evolution of the system can be followed for long period of times. These schemata are approximate and there are errors associated with them. In particular, we can have truncation and rounding off errors. Truncation errors are related to the accuracy of the finite difference method with respect to the true solution. Finite difference methods are usually based on a Taylor expansion truncated at some term. These errors are independent on the implementation. They are intrinsic to the algorithm. Round-off errors, related to errors associated to a particular implementation of the algorithm. It is based on finite number of digits used in computer arithmetics. Both errors can be reduced by decreasing Δt. For large Δt, truncation errors dominate, but they decrease quickly as Δt is decreased. Round-off errors decrease more slowly with decreasing Δt, and dominate in the small Δt limit. With 64-bit precision helps to keep round-off errors at a minimum level.

There are many different type of models that have been developed and tested to perform molecular dynamic simulations. They can be divided as unconstrained and constrained simulation schemata.

1.5 Unconstrained molecular dynamic simulations

In this section, we discussed some of the popular algorithms which do not use constraints. These include Verlet, Leap-Frog, Predictor-Corrector, Velocity Verlet.

1.5.1 Verlet algorithm

The verlet algorithm was introduced by (49). Even this simple finite difference scheme is widely used in molecular dynamic simulations. A differential equation of the form (1.7) is a second order strongly non linear ordinary differential equation. We assume $x(t)$ represent 3 dimension position vector and consider Tayler expansion as follows (47):

$$x_i(t + \Delta t) = x_i(t) + \Delta t \dot{x}_i + \frac{1}{2!}\Delta t^2 \ddot{x}_i + \frac{1}{3!}\Delta t^3 \dddot{x}_i + O(\Delta t^4) \tag{1.8}$$

$$x_i(t - \Delta t) = x_i(t) - \Delta t \dot{x}_i + \frac{1}{2!}\Delta t^2 \ddot{x}_i - \frac{1}{3!}\Delta t^3 \dddot{x}_i + O(\Delta t^4) \tag{1.9}$$

Adding (1.8) and (1.9) give:

$$x_i(t + \Delta t) \quad = \quad x_i(t) - x_i(t - \Delta t) + \Delta t^2 \ddot{x}_i + O(\Delta t^4) \tag{1.10}$$

This is the basic form of the Verlet algorithm. Since, we integrate Newton's equation, by (1.7), we have:

$$\ddot{x}_i \quad = \quad \frac{1}{m_i} f_i = -\frac{1}{m_i}\nabla\varphi_i(x(t)) \tag{1.11}$$

where $\varphi_i(x(t))$ is a potential function. In beginning of this Chapter, we discuss potential functions in details.

$$x_i(t + \Delta t) \quad = \quad x_i(t) - x_i(t - \Delta t) - \Delta t^2 \frac{1}{m_i}\nabla\varphi_i(x(t)) + O(\Delta t^4) \tag{1.12}$$

We need initial values $x_i(0)$ and $x_i(\Delta t)$ to calculate preceding position. By throwing out $O(\Delta t4)$ term, we obtain recursive explicit formula to compute $x_i(t+\Delta t)$, $x_i(t+2\Delta t)$.....$x_i(t+n\Delta t)$ successively. The scheme in equation (1.12) is called Verlet algorithm (49). The velocities do not participate in the recursive iteration but are needed for property calculations. This makes it difficult to implement stochastic collisions for the equilibration of the temperature and impossible to use this method to solve differential equation, such as those arising in the constant pressure method, in which the acceleration depend upon the velocities as well as the position. However, the velocity can be calculated by:

$$v_i(t) \quad = \quad \dot{x}_i(t) = \frac{1}{2\Delta t}\left[x_i(t+\Delta t)-x_i(t+\Delta t)\right]+O(\Delta t^2) \qquad (1.13)$$

The computed $x_i(t+\Delta t)$ would be off from the real $x_i(t+\Delta t)$ by $O(\Delta t^4)$. We called this as a local truncation error which is intrinsic property of the algorithm. Clearly, as $\Delta t \to 0$, then local truncation error $\to 0$, but that does not guarantee the algorithm works, because what we need is $\{x_n(t+T)\}$ for a given finite T, not $x_i(t+\Delta t)$. To obtain $\{x_n(t+T)\}$, we must integrate $n_T\left(\frac{T}{\Delta t}\right)$ steps. The difference between computed $\{x_n(t+T)\}$ and the real $\{x_n(t+T)\}$ is called the global error. An algorithm can be useful only when $\Delta t \to 0$ the global error $\to 0$. A careful analysis of the error propagation in equation (1.10) indicates that the global error is $O((\Delta t)^2)$ as $\Delta t \to 0$. The Verlet algorithm is thus a second order method. This implies only part of the analysis because the order of an algorithm only characterizes its performance when $\Delta t \to 0$. To save computational cost, most often we must adopt a quite large Δt. Higher order algorithms do not necessarily perform better than lower order algorithms at practical Δt. In fact, they could be much worse by diverging spuriously at a certain Δt, while a more robust method would just give a finite but manageable error for the same Δt. This is the concept of the stability of a numerical algorithm.

In addition to local truncation error, there is round off error due to the computers finite precision. The effect of round off error can be better understood in the stability domain. In most applications, the (round off error) $\ll$ (local truncation error). Some applications, especially those involving high order algorithms, do push the machine precision limit. In those cases, equating (local truncation error) $\gg \mathcal{E}$ where $\mathcal{E}$ is the machines relative accuracy, provides a practical lower bound to Δt, since by reducing Δt would no longer reduce the global error (47).

Algorithm:

Start with $x_i(t)$ and $x_i(t - \Delta t)$

Repeat following steps:

1. Calculate $\dfrac{f_i}{m_i}\left(= -\dfrac{\nabla \varphi_i}{m_i}\right)$

2. Calculate $x_i(t + \Delta t)$ using equation (1.12)

3. Calculate $v_i(t)$ if desired

4. Replace $x_i(t - \Delta t)$ with $x_i(t)$ and $x_i(t)$ with $x_i(t + \Delta t)$

5. Stop if it converges otherwise repeat step 1

Verlet algorithm is computed the advancement of positions all in one step using equation (1.12). It is simple to program since it is simple straight forward algorithm. Verlet scheme is time reversible and conserves energy well even with relatively long time steps. The velocities at t can be calculated only after $x_i(t + \Delta t)$ are known. One must know initial $x_i(t)$ and $x_i(t - \Delta t)$ to start trajectory, rather than $x_i(t)$ and $v_i(t)$ (2).

1.5.2 Leap-Frog algorithm

Leaf-frog (47) method is a modified version of the Verlet algorithm. As we describe in previous section, the Verlet algorithm uses the positions and force at the time t and the positions at the time $(t-\Delta t)$ to predict the positions at the time $(t+\Delta t)$, where Δt is the integration step. The error in the atomic positions is of the order of $O(\Delta t^4)$. The velocities are calculated from the basic definition of differentiation of equation (1.13) with an error of the order of $O(\Delta t^2)$. To obtain more accurate velocities, the leapfrog algorithm which uses velocities at half time step can be used:

$$v_i(t+\frac{\Delta t}{2}) \;=\; v_i(t-\frac{\Delta t}{2}) + \Delta t\,\frac{f_i}{m_i} + O\left((\Delta t)^3\right) \tag{1.14}$$

Atomic position calculate:s

$$x_i(t+\Delta t) \;=\; x_i(t) + \Delta t\,v_i(t+\frac{\Delta t}{2}) + O\left((\Delta t)^3\right) \tag{1.15}$$

Velocity at time t approximated:

$$v_i(t) \;=\; \frac{v_i(t-\frac{\Delta t}{2}) + v_i(t-\frac{\Delta t}{2})}{2} + O\left((\Delta t)^2\right) \tag{1.16}$$

This method is useful when the kinetic energy is needed at time t. The leapfrog algorithm is computationally less expensive and requires less storage which could be an important advantage in the case of large scale calculations. Moreover, the conservation of energy is respected, even at large time steps. Therefore, the computation time could be greatly decreased when this algorithm is used. However, when more accurate velocities and positions are needed, another algorithm should be implemented, such as Predictor-Corrector algorithm.

1.5.3 Predictor-Corrector algorithm

Here, we solve the second order differential equation (1.7). That can be written in normal form:

$$\ddot{x} = f(\dot{x}, x, t) \tag{1.17}$$

where $x \in \mathbb{R}^3$. First step of this algorithm consists in evaluating the atomic positions and velocities at time $(t + \Delta t)$ from the positions and the velocities at time $(t - i\Delta t)$, where $i = 0; \dots k - 2$. k is the order of the predictor part. The extrapolation is given by:

$$x_i(t + \Delta t) = x_i(t) + \Delta t\, \dot{x}_i(t) + \Delta t^2 \sum_{i=1}^{k-1} \alpha_i f\left(t + \Delta t(1 - i)\right) \tag{1.18}$$

which compute atomic position and:

$$\Delta t\, \dot{x}_i(t) = x_i(t + \Delta t) - x_i(t) + \Delta t^2 \sum_{i=1}^{k-1} \beta_i\, f\left(t + \Delta t(1 - i)\right) \tag{1.19}$$

for the velocities. The coefficient β_i *satisfy* the following equation:

$$\sum_{i=1}^{k-1} (1 - i)^q \beta_i = \frac{1}{q+2}, \quad q = 0, 1, \dots, k - 2 \tag{1.20}$$

The algorithm constitutes another commonly used class of method to integrate the equation of motion. Algorithm is contained three computational steps. They are:

1. Predictor : From the positions and their time derivatives up to a certain order q, all known at time t, one predict the same quantities at time $(t + \Delta t)$ by means of Taylor expansion.

2. Force evaluation: Force is computed taking the gradient of the potential at the predicted positions. Resulting acceleration will be in general different from the predicted acceleration. The difference between the two constitutes an error.

3. Corrector : Define an error used to correct positions and their derivatives. All the corrections are proportional to error. The coefficient of proportionality being a magic number determined to maximize the stability of the algorithm (47).

The Predictor Corrector algorithm gives more accurate positions and velocities than the leapfrog algorithm, and is therefore suitable in very delicate calculations. However, it is computationally expensive because it include additional step and needs significant storage.

1.5.4 Velocity version of Verlet algorithm

In the Verlet algorithm the velocities are not calculated explicitly and leads to difficulties in some applications. Because, the velocity of time t can be calculated only after the position at time $(t+\Delta t)$ has been obtained. Making it difficult to implement simulations such as constant pressure since it is depends on velocities as well as positions. The velocity Verlet algorithm (46) overcomes this difficulty:

$$x_i(t+\Delta t) = x_i(t) + \Delta t\, \dot{x}_i(t) + \frac{\Delta t^2}{2}\frac{f_i(x(t))}{m_i} + O(\Delta t^3) \tag{1.21}$$

$$\dot{x}_i(t+\Delta t) = \dot{x}_i(t) + \frac{\Delta t^2}{2}\left[f_i(x(t)) + f(x(t+\Delta t))\right] + O(\Delta t^2) \tag{1.22}$$

Algorithm:

Start with $x_i(t)$, $\dot{x}_i(t)$ and calculate $f_i(x(t))$. Repeat the following steps:

1. Calculate $x_i(t+\Delta x)$ using equation (1.21)

2. Calculate velocities at mid-step using $\dot{x}_i(t+\frac{\Delta t}{2}) = \dot{x}_i(t) + \frac{\Delta t}{2}f_i(x(t))$

3. Calculate $f_i(t+\Delta t)$

4. Compute the velocity using $\dot{x}_i(t+\Delta t) = \dot{x}_i(t+\frac{\Delta t}{2}) + \frac{\Delta t}{2}f_i(t+\Delta t)$

5. Stop if converge otherwise repeat step 1

This version of algorithm does calculate position and velocity simultaneously. Local and global errors are in order of $O((\Delta t)^3)$ and $O((\Delta t)^2)$ respectively. Since, Velocity version of Verlet algorithm calculate velocities and positions simultaneously, it enable us to compute kinetic energy at time $(t + \Delta t)$. Velocity version of Verlet algorithm is numerically stable, and can start with positions and velocities at time t. Studies have shown that the scheme conserves energy well even with relatively long time steps and simple to program.

1.5.5 Beemans algorithm

Beeman's model (6) is similar to the velocity Verlet algorithm. We start out with $x_i(t)$, $f_i(t - \Delta t), f_i(t)$ and $\dot{x}_i(t)$. Then:

$$x_i(t + \Delta t) = x_i(t) + \Delta t\, \dot{x}_i(t) + \frac{\Delta t^2}{6m_i}\left[4f_i(t) - f_i(t - \Delta t)\right] + O(\Delta t^4) \qquad (1.23)$$

evaluate $f_i(t + \Delta t)$ and then:

$$\dot{x}_i(t + \Delta t) = \dot{x}_i(t) + \frac{\Delta t}{12m_i}\left[5f_i(t + \Delta t) + 8f_i(t) - f_i(t - \Delta t)\right] + O(\Delta t^4) \qquad (1.24)$$

This is a third order method.

1.5.6 Symplectic integrators

Symplectic integrators preserve the property of phase space volume conservation (Liouvilles theorem) of Hamiltonian dynamics. They tend to have much better energy conservation in the long run. The velocity Verlet algorithm is, in fact, symplectic [(52), (44)]. As with the predictor-corrector algorithm, symplectic integrators tend to perform better at higher order, even on a per cost basis. The high-order predictor-corrector and high-order symplectic integrators are the real competitors for high accuracy integrators. It has been understood that the long term performance of a symplectic integrator is always superior to that of a non-symplectic integrator (47).

1.6 Constrained molecular dynamic simulations

A common modeling (4) strategy in molecular dynamics is to maintain atoms at fixed separations by the use of constraint relations in cartesian coordinates. This approach can be generalized to freeze other relationships among the other variables, as well. Constrained molecular dynamic methods are popular, especially to fixed intramolecular bond lengths and/or angles during a simulation. Intramolecular bond vibrations are typically the highest frequencies in the system and therefore determine the largest time step that can be used. If bonds are constrained, then a larger time step can be used, which speeds up the computation (23).

1.6.1 Shake algorithm

The Shake algorithm is introduce by (38). This is the procedure to integrate the equation of motion with internal constraints. It has been shown that when internal constraints are present, then the equation of motion can be written as:

$$m_i \ddot{x}_i(t) \;=\; f_i\big(x_i(t)\big) + C\big(x_i(t), \ddot{x}_i\big) \tag{1.25}$$

where $C(x)$ represents forces associated with the constraints. The force function C describes the mechanical state of the system. The nature of the constraints is dependent on the functional form of state of mechanical system as well. The Shake model can be formed as:

$$x_i(t + \Delta t) = 2x_i(t) - x_i(t - \Delta t) + \frac{\Delta t^2}{m_i}\Big[f_i\big(x_i(t)\big) + C_i\big(x_i(t), \dot{x}_i(t)\big) \Big] \tag{1.26}$$

where $x_i \in \mathbb{R}^3$. The major difficulty associated with equation (1.26) is that, even if we use exact function C, the intramolecular constraints would be violated due to the fact that the Shake algorithm is not exact. In 1977, Rychaert was shown that this can be overcome by not using exact function C but by using the approximation for C. The method requires $x(t + \Delta t)$ to satisfy constraints within a desired accuracy. This method can be derived by time dependent Lagrange multiplier method (Chapter 2). The important fact of this algorithm is that it has local error of order $O((\Delta t)^4)$. $x(t + \Delta t)$ can be computed by following iterative schemata (38):

$$x_i(t + \Delta t) = 2x_i(t) - x_i(t - \Delta t) + \frac{\Delta t^2}{m_i}\left[f_i\left(r_i(t)\right) + G_i(t)\right]$$

(1.27)

where G_i is a approximation to C_i. Iteration can not proceed unless we know $x_i(t)$ and $x_i(t - \Delta t)$. The Shake scheme has the same advantages and disadvantages like Verlet. To eliminate disadvantages of the schemata, the Rattle algorithm was introduced. The Rattle algorithm is also called the Velocity Version of the Shake algorithm.

1.6.2 Rattle algorithm

Anderson (2) was introduced following constraints schemata to calculate velocity and position simultaneously.

$$x_i(t + \Delta t) = x_i(t) + \Delta t\, \dot{x}_i(t) + \frac{\Delta t^2}{2m_i}\left[f_i\left(x_i(t)\right) + C\left(x_i(t), \dot{x}_i(t)\right)\right]$$

(1.28)

$$\dot{x}_i(t + \Delta t) = \dot{x}_i(t) + \frac{\Delta t}{2m_i}\left[f_i\left(x_i(t)\right) + C\left(x_i(t), \dot{x}_i(t)\right) + f_i\left(x_i(t + \Delta t)\right) + \right.$$

$$\left. C\left(x_i(t + \Delta t), \dot{x}_i(t + \Delta t)\right)\right]$$

(1.29)

where $x \in \mathbb{R}^3$. Like the Velocity Version of the Verlet, the position $x_i(t)$ and velocity $\dot{x}_i(t)$ of initial structure are required to start simulation. Then, we can calculate $x_i(t + \Delta t)$ by replacing $C(x_i(t), \dot{x}_i(t))$ by an approximation that made $x_i(t + \Delta t)$ satisfy the constraints.

In the equation (1.29) the term $\dot{x}_i(t + \Delta t)$ appears in the both side of the equation. This is inconsistent with a simple iterative schema and can be eliminated by using two different approximation function for equations (1.28) and (1.29). Thus, we have:

$$x_i(t + \Delta t) \;=\; x_i(t) + \Delta t \dot{x}_i(t) + \frac{\Delta t^2}{2m_i}\Big[f_i\big(r_i(t)\big) + G_i(t)\Big] \tag{1.30}$$

$$\dot{x}_i(t + \Delta t) \;=\; \dot{x}_i(t) + \frac{\Delta t}{2m_i}\Big[f_i\big(x_i(t)\big) + G_i(t) + f_i\big(x_i(t + \Delta t)\big) + H_i(t)\Big] \tag{1.31}$$

The Rattle algorithm makes two different approximations $G_i(t)$ and $H_i(t)$ for the forces associate with the constraints. It is possible to obtain both positions and velocities simultaneously which satisfy the constraints. It has been proved that (2), Rattle has a local error of order $((\Delta t)^3)$ and global error of order $((\Delta t)^2)$ which is same as in the velocity form of the Verlet algorithm for unconstrained dynamics.

Liouville equation is used especially for simulations with constant pressure and constant temperature ensembles. It describes the evolution of the phase space distribution function for the conservative (4) Hamiltonian system which continuity equation for the flux. Therefore, we derived Liouville equation in following section.

1.6.2.1 Liouville equation

Construct a cartesian space in which each of the $6n$ coordinates and momenta is assigned to one of $6n$ mutually orthogonal axes. Phase space is a $6n$ dimensional space (47). A point in this space is specified by giving a particular set of values for the $6n$ coordinates and momenta. Denote such a point by $x = (p_1, ..., p_n, q_1, ...q_n)$. x is a $6n$ dimensional vector. Thus, the time evolution or trajectory of a system as specified by Hamilton's equations of motion, can be expressed by giving the phase space vector, as a function of time. The law of conservation of energy, expressed as a condition on the phase space vector $H(x(t)) = \varphi =$ constant defines a $(6n - 1)$ dimensional hyper-surface in phase space on which the trajectory must remain.

Consider phase space for the ensemble of the n-particle systems. The number of systems in the ensemble is constant. We can write the continuity equation for density $\rho(p_1, \ldots, p_n, q_1, \ldots q_n)$ in phase space:

$$\frac{\partial \rho}{\partial t} + div(\rho v) = 0 \tag{1.32}$$

For the flux in phase space, coordinates are p's and q's and velocities are p's and q's:

$$\frac{\partial \rho}{\partial t} + \sum_{i=1}^{n}\left[\frac{\partial(\rho \dot{q}_i)}{\partial q_i} + \frac{\partial(\rho \dot{p}_i)}{\partial p_i}\right] = 0 \tag{1.33}$$

Applying chain rule, we obtain:

$$\frac{\partial \rho}{\partial t} + \sum_{i=1}^{n}\left[\dot{q}\frac{\partial(\rho)}{\partial q_i} + \dot{p}\frac{\partial(\rho)}{\partial p_i}\right] + \rho\sum_{i=1}^{n}\left[\frac{\partial(\dot{q})}{\partial q_i} + \dot{p}\frac{\partial(\dot{p})}{\partial p_i}\right] = 0 \tag{1.34}$$

Using Hamilton equations of motion:

$$\dot{q}_i = \frac{\partial H}{\partial p_i} \quad \dot{p}_i = -\frac{\partial H}{\partial q_i} \tag{1.35}$$

Thus:

$$\frac{\partial \dot{q}_i}{\partial q_i} = \frac{\partial^2 H}{\partial p_i \partial q_i} = \frac{\partial \dot{p}_i}{\partial p_i} \tag{1.36}$$

Thus the second sum in equation (1.34) is zero, and:

$$\frac{\partial \rho}{\partial t} + \sum_{i=1}^{n}\left[\dot{q}\frac{\partial(\rho)}{\partial q_i} + \dot{p}\frac{\partial(\rho)}{\partial p_i}\right] = 0 \tag{1.37}$$

The equation (1.37) is called Liouville equation. The left hand side of equation (1.37) is actually the full derivative of the distribution function, describing its change along the trajectories (47). In conservative systems the distribution function is constant along the trajectories, being an integral of motion.

1.6.3 Stochastic method

Coupling to the environment is simulated by random collisions with imaginary heat bath particles. These collisions lead to instantaneous momentum changes. The particle momenta are reset to new values, taken from the Maxwell distribution. This way the average kinetic energy is always correct. The natural variation on this theme is resetting velocities of all particles at the same time after certain interval. Then the dynamics during this interval is truly microcanonical, and time correlation functions can be calculated inside this interval. After the new velocities are assigned, the new con_guration is accepted or rejected based on Metropolis-like criteria for Monte Carlo simulations.

This technique was first suggested by Heyes in 1980, and then reinvented by D. Heermann et. al. in 1990. Hybrid Monte Carlo. D. Heermann et. al. showed, that this acceptance criterium is needed to account for the numerical integration errors. Otherwise this technique reproduces canonical ensemble only approximately.

1.6.4 Velocity rescaling

An alternative way to simulate constant temperature is to re-scale all the velocities to keep kinetic energy constant. It is a very unrealistic approach used in the early days. If done on every step, it alters the system dynamics, which does not even correspond to the canonical ensemble. If done at certain intervals, it adds some periodic perturbation to the system, which is in general undesirable, but sometimes can serve as a tool to study system dynamics. This was used for simulations of glasses by Rahman et. al. in early 1980's. It is also often used to equilibrate the system during the the first few hundred MD steps before the production run starts and data are collected.

A more gentle and more practical way, known as Van Gunstern-Berendsen thermostat is to use a factor, that depends on the deviation of the instantaneous kinetic energy from the average value, corresponding to desired temperature. At each time step velocities are scaled by the some factor is the molecular dynamic time step, and is a parameter, that defines, how strong is the thermostat inuence.

1.7 Review

In this section, constrained and unconstrained molecular dynamic simulation methods are described. Previous studies (4) have proved that the freezing high vibration frequency motions of atoms not badly effect on physical characteristics of the atomic systems. However, It enables to increase size of the time step without altering the system properties. Furthermore, some of the methods have been modified to run on parallel computers with many processors which can speed up computation as well as without introducing significant round off errors.

CHAPTER 2.
LAGRANGE MULTIPLIER METHOD

Lagrange multiplier method is used to solve a system of equations with constrains. This method is widely used and very popular in many research fields. In this chapter, the Lagrange multiplier method and time dependent Lagrange multiplier method are discussed in details. This technique has applied to molecular dynamic simulation in 1988 (48). When used to solve non linear system, this method is computationally expensive. The Lagrange multiplier method is used when we need to find the extreme values of a function whose domain is constrained to lie within a particular subset of the domain. The Lagrange multiplier rule was introduced in 1762 (27). In 1788, Lagrange proved that it can be used for minimizing a function subject to equality constraints.

2.0.1 Lagrange multiplier method

In physics and engineering problems (8) we may be called upon to find the maxima or minima of a function of several variables:

$$\min [\, f(x) \,]$$

$$\text{subject to } C(x) = 0$$

(2.1)

where x is a multi-variable function. $f(x)$ and $C(x)$ are objective and constraints functions respectively. Lagrange method consists of introducing a new function which incorporates $f(x)$ together with all the constraints. The new function is called a Lagrangian. Therefore, the Lagrangian is written as:

$$\mathbb{L} \;=\; f(x) - \lambda C(x)$$

(2.2)

where λ is a constant called a Lagrange multiplier. The conditions are defined:

$$\nabla \mathbb{L} = \nabla\big(f(x) - \lambda C(x)\big) = 0$$
$$C(x) = 0$$

(2.3)

Here, we need to determine;

- Lagrange multiplier constant λ
- Extremal values of x
- The value min $[\,f(x)\,]$

The advantage in this approach is that it treats all variables and constraints in a symmetric fashion so that problems involving many variables and constraints can be neatly organized. Depend on $f(x)$ and $C(x)$, it is necessary to solve linear or non-linear system of equations to determine unknowns (34).

2.0.2 Time dependent Lagrange multiplier method for molecular dynamics

The system represent in equation (2.3) is called time dependent Lagrange multiplier method if parameter λ is a function of time t. Assume that a molecular system with n atoms and they are interacting via a potential energy $\varphi(r)$. Thus, one could introduce constraints (48):

$$C_k\big(x_j(t)\big) = 0 \quad k = 1,\dots m \quad \text{and} \quad j = 1,\dots n$$

(2.4)

We define Lagrange equation of motion:

$$m_i \ddot{r} = f_i + C_i \tag{2.5}$$

where C_i represents total constraints force acting on atom i. Therefore, we have:

$$m_i \ddot{x} = -\nabla \varphi(x_i(t)) - \sum_{k=1}^{m} \lambda_k(t) \nabla C_k(x_i(t)) \tag{2.6}$$

where the m represents number of constraints in the system. The λ_k Lagrange multipliers are time dependent and determined by requiring that, the constraints in equation (2.4) are satisfied exactly. The equation (2.6) and (2.4) generate ($3n$) and (m) number of equations respectively. The system has ($3n + m$) number of equations to find ($3n + m$) number of unknowns. Therefore, we can uniquely determine the unknowns provided it is consistent. In general, the numerical integration methods are used to obtain the trajectory. It has been shown that, it is not convenient to find λ_k's in terms of positions and velocities. Without loss of accuracy, we assume that λ_k's are parameters. Finite difference schema can be used to calculate next position as follows:

$$x_i(t+\Delta t) = 2x_i(t) - x_i(t-\Delta t) - \frac{\Delta t^2}{m_i} \left[\nabla \varphi(x_i(t)) + \sum_{k=1}^{m} \lambda_k(t) \nabla C_k(x_i(t)) \right] \tag{2.7}$$

This generates positions at time ($t + \Delta t$) satisfy constraints in (2.4). That can be written formally as:

$$C_k(x_j(t+\Delta t)) = 0 \quad k = 1, \ldots m \quad \text{and} \quad j = 1, \ldots n \tag{2.8}$$

Let's divide equation (2.4) into two so that they represent constraint and unconstraint motion.

$$x_i(t+\Delta t) = p_i(t+\Delta t) + q_i(t+\Delta t) \tag{2.9}$$

where:

$$p_i(t+\Delta t) = 2x_i(t) - x_i(t-\Delta t) - \frac{\Delta t^2}{m_i} \nabla \varphi(x_i(t)) \tag{2.10}$$

$$q_i(t+\Delta t) = -\frac{\Delta t^2}{m_i} \sum_{k=1}^{m} \lambda_k(t) \nabla C_k(r_i(t)) \tag{2.11}$$

Solutions to the motion (48) given by equations (2.9) to (2.11) are exact third order in the time step (Δt). It is obvious to see that the schema has same accuracy (38) that of Verlet (equation 1.12). To keep the arbitrary internal degree of freedom fixed during the simulation, the following form of harmonic constraints has been chosen:

$$C_k\left(x_j(t)\right) \;=\; \delta d_k = d_k(t+\Delta t) - d_k(t) \tag{2.12}$$

where $d_k(t+\Delta t)$ and $d_k(t)$ are arbitrary internal coordinates at time $(t+\Delta t)$ and t respectively. The δd_k represents internal coordinates variations over time steps Δt. We need equation (2.12) to be zero in order to get constraints satisfied. For our convenient, we introduce $d_k = d_k(x_1, \ldots, x_n)$ where n is the number of atoms defining the d_k. The small changes of Cartesian coordinates produces changes of δd_k (total differential):

$$\delta d_k \;=\; \sum_{j=1}^{n}\left[\frac{\partial d_k}{\partial x_j}\bar{i} + \frac{\partial d_k}{\partial y_j}\bar{j} + \frac{\partial d_k}{\partial z_j}\bar{k}\right]\left[x_j(t+\Delta t) - x_j(t)\right] \tag{2.13}$$

where $\bar{i}$, $\bar{j}$ and $\bar{k}$ are unit vectors along the usual x, y and z axis. Define d_{kj} such that:

$$\delta d_k \;=\; \sum_{j=1}^{n}\alpha_{kj}[x_j(t+\Delta t) - x_j(t)] \tag{2.14}$$

$$\alpha_{kj} \;=\; \left[\frac{\partial d_k}{\partial x_j}\bar{i} + \frac{\partial d_k}{\partial y_j}\bar{j} + \frac{\partial d_k}{\partial z_j}\bar{k}\right] \tag{2.15}$$

If all the constraints are satisfied then $C_k(x_j(t)) = 0$. Therefore:

$$\sum_{j=1}^{n}\alpha_{kj}\left[x_j(t+\Delta t) - x_j(t)\right] \;=\; 0 \tag{2.16}$$

substituting $x_j(t+\Delta t)$ by equation (2.9);

$$\sum_{j=1}^{n}\alpha_{kj}\left[p_j(t+\Delta t) + q_j(t+\Delta t) - x_j(t)\right] \;=\; 0 \tag{2.17}$$

$$\sum_{j=1}^{n}\alpha_{kj}q_j(t+\Delta t) \;=\; \sum_{j=1}^{n}\alpha_{kj}\left[x_j(t) - q_j(t+\Delta t)\right] \tag{2.18}$$

The equation (2.11) gives:

$$q_i(t+\Delta t) \;=\; -\frac{\Delta t^2}{m_i}\sum_{k=1}^{m}\lambda_k(t)\nabla[\delta d_k] \qquad (2.19)$$

$$q_i(t+\Delta t) \;=\; -\frac{\Delta t^2}{m_i}\sum_{k=1}^{m}\lambda_k(t)\alpha_{ki} \qquad (2.20)$$

replacing $q_i(t+\Delta t)$ in equation (2.18):

$$\sum_{s=1}^{m}\sum_{j=1}^{n}\frac{1}{m_j}\alpha_{kj}\,\alpha_{sj}\,\lambda_k(t) \;=\; \frac{-1}{\Delta t^2}\sum_{j=1}^{n}\alpha_{kj}\Big[x_j(t)-q_j(t+\Delta t)\Big] \qquad (2.21)$$

This could be written in matrix form for non singular D:

$$D\,\Lambda \;=\; \frac{-1}{\Delta t^2}\Omega \qquad (2.22)$$

$$\Lambda \;=\; \frac{-1}{\Delta t^2}D^{-1}\Omega \qquad (2.23)$$

Here D, Λ and Ω are $(m \times m)$, $(m \times m)$ and $(m \times m)$ matrices. It has been shown (48) that the matrix D is a Wilson matrix and therefore it is non-singular. The equation (2.20) might be rewritten in matrix form (38):

$$\Upsilon \;=\; -\Delta t^2 M^{-1}Q\Lambda \qquad (2.24)$$

Then by substituting equation (2.23) with (2.22):

$$\Upsilon \;=\; M^{-1}QD^{-1}\Omega \qquad (2.25)$$

where M is a $(diag(m_1,m_2, \ldots,m_n))$ diagonal matrix of particle masses with $det(M) \neq 0$. The matrix Q is a $(3n \times m)$ of the components of the α_{kj}.

2.1 Review
. . . .

The implementation of method begins by calculating $p_i(t+\Delta t)$ using equation (2.9). Thereafter, the equation (2.11) can be used to compute constraints part of the simulation to calculate Lagrange multipliers λk's. By using the previous two facts one can calculate the next position on the molecular atoms. This process can be iterated until it achieves desired accuracy. This procedure has the qualities to be used in any constrained molecular dynamic algorithm. The Lagrange multiplier algorithm is straightforward to implement. The technically difficult and time consuming part of the algorithm is solving non-linear constraints equations. This work has been presented in (48) paper.

CHAPTER 3.
PENALTY AND BARRIER METHODS

Penalty function methods are developed to eliminate some or all of the constraints and add to the objective function a penalty term which prescribes a high cost to infeasible points. In theory, penalty function method uses unconstraint optimization methods to solve constraints optimization problems. Discrete iterative setup can be started with infeasible or feasible starting point and guide system to feasibility and ultimately obtained optimal solution.

3.0.1 History

In 1943, Courant introduced the quadratic penalty method where the penalty term is the squared Euclidean norm of the constraint violations (11). In 1970, Fletcher is studied the Lagrange function depending only on the variables (17), then, gave the theoretical justification of a class of exact penalty methods for solving smooth equality constrained nonlinear optimization problems. Exact penalty methods (15) were intensively investigated and a well-prepared survey was published by Di Pillo (1994). A new smooth exact penalty (10) function was suggested by Christianson (1995). The Lagrange multiplier rule was further developed by Rapcsak [(40), (41)] who combined the optimization theory with Riemannian geometry in order to describe the geometric structure of smooth nonlinear optimization problems by tensors and to extend the local results of Lagrange to global ones. In (42), the idea of Fletcher (1970) to define smooth exact penalty functions and that of Courant (1943) to use a quadratic penalty term were reconsidered and developed further by the global version of the global Lagrange multiplier method clarifying the geometric meaning as well.

3.0.2 Constraints

Most optimization problems have constraints. The solution or set of solutions which are obtained as the final result of an evolutionary search must necessarily be feasible, that is, satisfy all constraints. A taxonomy of constraints can be considered and composed of a number, metric, criticality and difficulty. A first aspect is number of constraints, ranging upwards from one. Sometimes problems with multiple objectives are reformulated with some of the objectives acting as constraints. Difficulty in satisfying constraints will increase with the number of constraints. A second aspect of constraints is their metric, either continuous or discrete, so that a violation of the constraint can be assessed in distance from satisfaction using that metric. A third consideration is the criticality of the constraint, in terms of absolute satisfaction. A constraint is generally formulated as hard when in fact, it is often somewhat soft. That is, small violations would be considered for the final solution if the solution is superior to other solutions. Evolutionary algorithms are especially capable of handling soft constraints since a population of solutions is returned at each point in the search. This allows the user to select a solution which violates a soft constraint (infeasible) over a solution which would be the best, technically feasible solution found. A final aspect of constraints to be considered is the difficulty of satisfying the constraints. This difficulty can be characterized by the size of the feasible region compared to the size of the sample space. The difficulty may not be known a priori, but can be gauged in two ways. The first way is how simple it is to change a solution which violates the constraint to a solution but does not violate the constraint. The second way is the probability of violating the constraint during search. For example, a constraint may be frequently violated but the solution can be easily made feasible. Conversely, a constraint violation may be very difficult to resolve, but occur rarely in the search.

3.0.3 Penalty function method

Penalty functions have been a part of the literature on constrained optimization for decades (34). Three type of penalty functions are exist (5). They are called Barrier methods, partial penalty functions and global penalty functions (45). In general, a penalty function approach is as follows. Consider the constrained optimization problem:

$$\begin{aligned} \min \quad & \left(f(x)\right) \\ \text{such that} \quad & g_i(x) \;\leq\; 0 \quad i = 1, \ldots, m \\ & h_i(x) \;\leq\; 0 \quad i = 1, \ldots, l \\ \text{where} \quad & x \in \mathbb{R}^n \end{aligned}$$

(3.1)

whose feasible region we denote by $\Omega = \{x \in \mathbb{R}^n \mid g_i(x) \leq 0 \ i = 1, \ldots, m, \ h_i(x) = 0 \ i = 1, \ldots, l\}$ and write $g(x) = (g_1(x), \ldots, g_m(x))^T$ and $h(x) = (h_1(x), \ldots, h_l(x))^T$ for convenience.

Penalty methods are designed to solve (3.1) by, instead, solving a sequence of specially constructed unconstrained optimization problems (34). The feasible region of equation (3.1) is expanded from Ω to all of $\mathbb{R}^n$, but a penalty is added to the objective function for points that lie outside of the original feasible region Ω.

Definition 3.0.1. A function $C(x) : \mathbb{R}^n \to \mathbb{R}$ is called a penalty function for equation (3.1) if $C(x)$ satisfies: $\{C(x) = 0 \ \text{if} \ g(x) \leq 0 \ h(x) = 0\}$ and $\{C(x) > 0 \ \text{if} \ g(x) > 0 \ \text{or} \ h(x) \neq 0\}$ (5)

Penalty functions are typically defined by:

$$C(x) \; = \; \sum_{i=1}^{m} \phi\left(g_i(x)\right) + \sum_{i=1}^{l} \psi\left(h_i(x)\right) \tag{3.2}$$

where

$$\left\{\phi\left(g_i(x)\right)=0 \;\; \text{if} \;\; g_i(x) \leq 0\right\} \;\; \text{and} \;\; \left\{\phi\left(g_i(x)\right)>0 \;\; \text{if} \;\; g_i(x) > 0\right\},$$

$$\left\{\psi\left(h_i(x)\right)=0 \;\; \text{if} \;\; h_i = 0\right\} \;\; \text{and} \;\; \left\{\psi\left(h_i(x)\right)>0 \;\; \text{if} \;\; h_i \neq 0\right\}$$

In theory, more general functions satisfying the definition can conceptually be used. We then consider solving the following penalty program:

$$min\left(f(x)+\mu C(x)\right) \tag{3.3}$$
$$\text{where} \quad x \in \mathbb{R}^n$$

for an increasing sequence of constants μ as $\mu \to \infty$. In problem (3.3), we are assigning a penalty to the violated constraints. The scalar quantity μ is called the penalty parameter.

Let $\left\{\mu_k\right\}_{k=1}^{\infty}$ be an increasing sequence of penalty parameters that satisfies $\mu_{k+1} > \mu_k$ for $\forall k$ and $\lim_{k \to \infty}\left(\mu_k\right) \to +\infty$. Let $F(x) = f(x) + \mu C(x)$ and let x^k be the exact solution to the problem (3.3), and let x^* denote any optimal solution of (3.1). The following Lemma presents some basic properties of penalty methods.

Lemma 3.0.1. *Properties of penalty methods (5)*

1. $F\left(\mu_k, x^k\right) \leq F\left(\mu_k+1, x^{k+1}\right)$

2. $C\left(x^k\right) \geq C\left(x^{k+1}\right)$

3. $f(x^k) \leq f(x^{k+1})$

4. $f(x^*) \geq F\left(\mu_k, x^k\right) \geq f(x^k)$

proof:

1.

$$
\begin{aligned}
F(\mu_{k+1}, x^{k+1}) &= f(x^{k+1}) + \mu_{k+1} C(x^{k+1}) \\
&\geq f(x^{k+1}) + \mu_k C(x^{k+1}) \quad \text{since} \quad \mu_{k+1} > \mu_k \\
&\geq f(x^k) + \mu_k C(x^k) \\
&= F(\mu_k, x^k)
\end{aligned}
$$

2.

$$ f(x^k) + \mu_k C(x^k) \leq f(x^{k+1}) \mu_k C(x^{k+1}) $$

and $$ f(x^{k+1}) + \mu_{k+1} C(x^{k+1}) \leq f(x^k) + \mu_{k+1} C(x^k) $$

Thus $$ (\mu_{k+1} - \mu_k) C(x^k) \geq (\mu_{k+1} - \mu_k) C(x^{k+1}) $$

whereby $$ C(x^k) \geq C(x^{k+1}) $$

3. From the proof of part (1):

$$ f(x^{k+1}) + \mu_k C(x^{k+1}) \geq f(x^k) + \mu_k C(x^k) $$

But $$ C(x^k) \geq C(x^{k+1}) $$

which implies that $$ f(x^{k+1}) \geq f(x^k) $$

4.

$$
\begin{aligned}
f(x^k) &\leq f(x^k) + \mu_k C(x^k) \\
&\leq f(x^*) + \mu_k C(x^*) \\
&= f(x^*)
\end{aligned}
$$

The convergence of the penalty method can be discussed with the following theorem:

Theorem 3.0.2. *(Penalty Convergence Theorem) Suppose that $\Omega \neq 0$ and $f(x)$, $g(x)$, $h(x)$, and $C(x)$ are continuous functions. Let $\{x^k\}_{k=1}^{\infty}$, be a sequence of solutions to equation (3.3), and suppose the sequence $\{x^k\}_{k=1}^{\infty}$ is contained in a compact set. Then any limit point $\bar{x}$ of $\{x^k\}_{k=1}^{\infty}$ solves (3.1) (5).*

proof:

Let $\bar{x}$ be a limit point of $\{x^k\}_{k=1}^{\infty}$. From the continuity of the functions involved, $\lim_{k \to \infty} x^k = f(\bar{x})$

Also, from the Lemma 3.1.1:

$$F^* = \lim_{k \to \infty} F(\mu_k, x^k) \leq f(x^*)$$

$$(3.4)$$

Thus

$$\lim_{k \to \infty} \mu_k C(x^k) = \lim_{k \to \infty} \left[F(\mu_k, x^k) - f(x^*) \right]$$

$$(3.5)$$

$$= F^* - f(\bar{x})$$

$$(3.6)$$

But $(\mu \to \infty)$, which implies from the above that:

$$\lim_{k \to \infty} C(x^k) \to 0$$

$$(3.7)$$

Therefore, from the continuity of $C(x)$, $g(x)$ and $h(x)$, $C(\bar{x}) = 0$, and so $g(\bar{x}) \leq 0$ and $h(\bar{x}) = 0$, that is, $\bar{x}$ is a feasible solution of (3.1). Finally, from the Lemma 3.1.1, $f(x^k) \leq f(x^*)$ for all k, and so $f(\bar{x}) \leq f(x^*)$, which implies that $\bar{x}$ is an optimal solution of (3.1).

An often used class of penalty functions is:

$$C(x) = \sum_{i=1}^{m_1} max\left[0, g_i(x)\right]^c + \sum_{i=1}^{m_2} \left| h_i(x) \right|^c \quad \text{where} \quad c \geq 1, \quad m = m_1 + m_2$$

$$(3.8)$$

If $c = 1$, $C(x)$ in equation (3.8) is called the linear penalty function. This function may not be differentiable at points where $g_i(x) = 0$ or $h_i(x) = 0$ for some i. Setting $c = 2$ is the most common form of (3.8) that is used in practice, and is called the quadratic penalty function.

3.0.4 Karush-Kuhn-Tucker multipliers

Suppose the penalty function $C(x)$ is defined as (3.2). $C(x)$ might not be continuously differentiable, since the functions $g_i(x)$ are not differentiable at points x where $g_i(x) = 0$. However, if we assume that the functions $\phi(y)$ and $\varphi(y)$ are continuously differentiable and $\phi'(0) = 0$ then $C(x)$ is differentiable whenever the functions $g(x)$, and $h(x)$ are differentiable, and we can write:

$$\nabla C(x) \;=\; \sum_{i=1}^{m} \phi'\big(g_i(x)\big)\nabla g_i(x) + \sum_{i=1}^{l} \varphi'\big(h_i(x)\big)\nabla h_i(x) \tag{3.9}$$

Let x_k solve (3.3). Then x^k will satisfy:

$$\nabla f(x^k) + \mu_k \nabla C(x^k) \;=\; 0 \tag{3.10}$$

that is:

$$\nabla f(x^k) + \mu_k \left[\sum_{i=1}^{m} \phi'\big(g_i(x^k)\big)\nabla g_i(x^k) + \sum_{i=1}^{l} \varphi'\big(h_i(x^k)\big)\nabla h_i(x^k) \right] \;=\; 0 \tag{3.11}$$

Define:

$$
\begin{aligned}
u_i^k &= \mu_k \phi'\big(g_i(x^k)\big) \\
v_i^k &= \mu_k \varphi'\big(h_i(x^k)\big)
\end{aligned}
\tag{3.12}
$$

$$\nabla f(x^k) + \sum_{i=1}^{m} u_i^k \nabla g_i(x^k) + \sum_{i=1}^{l} v_i^k \nabla h_i(x^k) \;=\; 0 \tag{3.13}$$

The u_i^k and v_i^k are called Karush-Kuhn-Tucker multipliers.

Lemma 3.0.3. *Suppose $\phi(y)$ and $\varphi(y)$ are continuously differentiable and satisfy $\phi(0) = 0$, and that $f(x)$, $g(x)$, and $h(x)$ are differentiable. Let u^k, v^k be defined by equation (3.12).*

Then if $x^k \to \bar{x}$, and $\bar{x}$ satisfies the linear independence condition for gradient vectors of active constraints, then $u^k \to u^k \to \bar{u}$ where $\bar{u}$ and $\bar{v}$ are a vector of Karush-Kuhn-Tucker multipliers for the optimal solution $\bar{x}$ of (3.1) (5).

proof:
From the Penalty Convergence Theorem (3.1.2), $\bar{x}$ is an optimal solution of (3.1). Let $I = \{i \mid g_i(\bar{x}) = 0\}$ and $N = \{i \mid g_i(\bar{x}) < 0\}$. *For* $i \in N$, $g_i(x^k) < 0$ for all k sufficiently large, so $u^k_i = 0$ for all k sufficiently large, whereby $\bar{u}_i = 0$ for $i \in N$. From equation (3.12) and the definition of a penalty function, it follows that $u^k_i \geq 0$ for $i \in I$, for all k sufficiently large. Suppose $u^k \to \bar{u}$ and $v^k \to \bar{v}$ as $k \to 1$. Then $\bar{u}_i = 0$ for $i \in N$. From the continuity of all functions involved:

$$\nabla f(x^k) + \sum_{i=1}^{m} u^k_i \nabla g_i(x^k) + \sum_{i=1}^{l} v^k_i \nabla h_i(x^k) = 0 \tag{3.14}$$

implies:

$$\nabla f(\bar{x}) + \sum_{i=1}^{m} \bar{u}_i \nabla g_i(\bar{x}) + \sum_{i=1}^{l} \bar{v}_i \nabla h_i(\bar{x}) = 0 \tag{3.15}$$

We also have $\bar{u} \geq 0$ and $\bar{u}_i = 0$ for all $i \in N$. Thus $\bar{u}$ and $\bar{v}$ are Karush-Kuhn-Tucker multipliers. It therefore remains to show that $u_k \to \bar{u}$ and $v_k \to \bar{v}$ for some unique $\bar{u}$ and $\bar{v}$.

3.0.5 Exact penalty function

The idea in an exact penalty method is to choose a penalty function $C(x)$ and constant μ so that the optimal solution $\bar{x}$ of (3.3) is also an optimal solution of the original problem in equation (3.1).

Theorem 3.0.4. *Suppose 3.1 is a convex program for which the Karush-Kuhn-Tucker conditions are necessary. Suppose that*

$$C(x) = \sum_{i=1}^{m} \left(g_i(x) \right)$$

Then as long as μ is chosen sufficiently large, the sets of optimal solutions of $F(\mu)$ and 3.1 coincide. In fact, it suffices to choose $\mu > \max_i \{u^*_i\}$, where u^* is a vector of Karush-Kuhn-Tucker multipliers.

Proof: Suppose $\bar{x}$ solves 3.1. For any $x \in \mathbb{R}^n$ we have:

$$F(\mu, x) = f(x) + \mu \sum_{i=1}^{m} g_i(x) \geq f(x) + \sum_{i=1}^{m} u_i^* g_i(x)$$

$$\geq f(x) + \sum_{i=1}^{m} u_i^* \left(g_i(\bar{x} + \nabla g(\bar{x})^T (x - \bar{x})) \right)$$

$$= f(x) + \sum_{i=1}^{m} u_i^* \nabla g(\bar{x})^T (x - \bar{x})$$

$$= f(x) - \nabla f(\bar{x})^T (x - \bar{x})$$

$$\geq f(\bar{x})$$

$$= f(\bar{x}) + \mu \sum_{i=1}^{m} g_i(\bar{x}) =$$

$$F(\mu, \bar{x})$$

Thus $F(\mu, \bar{x}) \leq F(\mu, x)$ for all x, and therefore $\bar{x}$ solves 3.3. Next suppose that $\hat{x}$ solves 3.3.

Then if $\hat{x}$ solves 3.1, we have:

$$f(\hat{x}) + \mu \sum_{i=1}^{m} g_i(\hat{x}) \leq f(\bar{x}) + \mu \sum_{i=1}^{m} g_i(\bar{x}) = f(\bar{x}) \quad \text{and}$$

$$f(\hat{x}) \leq f(\bar{x}) - \mu \sum_{i=1}^{m} g_i(\hat{x})$$

(3.16)

However, if $\hat{x}$ is not feasible for 3.1, then:

$$f(\hat{x}) \geq f(\bar{x}) + \nabla f(\bar{x})^T (\hat{x} - \bar{x})$$

$$= f(\bar{x}) - \sum_{i=1}^{m} u_i^* \nabla g_i(\bar{x})^T (\hat{x} - \bar{x})$$

$$\geq f(\bar{x}) + \sum_{i=1}^{m} u_i^* g_i(\bar{x}) - g_i(\hat{x})$$

$$= f(\bar{x}) - \sum_{i=1}^{m} u_i^* g_i(\hat{x}) > f(\bar{x}) - \mu \sum_{i=1}^{m} g_i(\hat{x})$$

which contradicts 3.16. Thus $\hat{x}$ is feasible for 3.1 and so $\hat{x}$ solves 3.1.

3.0.6 Barrier method

The idea in a barrier method is to dissuade points x from ever approaching the boundary of the feasible region (22).

Definition 3.0.2. A barrier function for 3.1 is any function $b(x): \mathbb{R}^n \to \mathbb{R}$ that satisfies, $b(x) = 0$ for all x that satisfy $g(x) < 0$, and $b(x) \to \infty$ as $\lim_{x} \max_{i} g_i(x) = 0$

Define barrier minimization problem:

$$\min \quad (f(x) + \mu b(x))$$
$$\text{s.t.} \quad g(x) \quad < \quad 0, \quad \text{for} \quad x \in \mathbb{R}^n \tag{3.17}$$

for a sequence of $\mu_k \to \infty$. The following Lemma presents some basic properties of barrier methods.

Lemma 3.0.5. *Let* $F(\mu, x) = f(x) + \mu_k b(x)$. *Let the sequence* $\{\mu_k\}$ *satisfy* $\mu_{k+1} > \mu_k$, $\mu_k \to \infty$ *as* $k \to \infty$. *Let* x^k *denote the exact solution to 3.17.*

- $F(\mu_k, x^k) \geq F(\mu_{k+1}, x^{k+1})$
- $b(x^k) \leq b(x^{k+1})$
- $f(x^k) \geq f(x^{k+1})$
- $f(x^*) \leq f(x^k) \leq F(\mu_k, x^k)$

Proof:

- $$\begin{aligned} F(\mu_k, x^k) &= f(x^k) + \mu_k b(x^k) \\ &\geq f(x^k) + \mu_{k+1} b(x^k) \\ &\geq f(x^{k+1}) + \mu_{k+1} b(x^{k+1}) \\ &= F(\mu_{k+1}, x^{k+1}) \end{aligned}$$

- $$f(x^k) + \mu_k b(x^k)) \leq f(x^{k+1}) + \mu_k b(x^{k+1})$$
  and $$f(x^{k+1}) + \mu_{k+1} b(x^{k+1})) \leq f(x^k) + \mu_{k+1} b(x^k)$$
  we have $$(\mu_k - \mu_{k+1}) b(x^k) \leq (\mu_k - \mu_{k+1}) b(x^{k+1})$$
  since $$\mu_k \leq \mu_{k+1}, \rightarrow b(x^{k+1}) > b(x^k)$$

  (3.18)

- From above proof, we have:

$$f(x^k) + \mu_{k+1} b(x^k) \geq f(x^{k+1}) + \mu_{k+1} b(x^{k+1}) \quad \text{and}$$
$$b(x^{k+1}) \geq b(x^k)$$

Therefore, $f(x^k) \geq f(x^{k+1})$

- $f(x^*) = f(x^k) = f(x^k) + \mu_k b(x^k) = F(\mu_k, x_k)$

Theorem 3.0.6. *Suppose $f(x)$, $g(x)$, and $b(x)$ are continuous functions. Let x^k, $k = 1,...$, be a sequence of solutions of $B(\mu_k)$. Suppose there exists an optimal solution x^* of 3.1 for which $N(\in, x^*) \cap \{x | g(x) < 0\} \neq \emptyset$ for every $\in > 0$. Then any limit point $\bar{x}$ of $\{x^k\}$ solves 3.1.*

proof: Let $\bar{x}$ be any limit point of the sequence $\{x^k\}$. From the continuity $f(x)$ and $g(x)$, $\lim_{k\to\infty} f(x^k) = f(\bar{x})$ and $\lim_{k\to\infty} g(x^k) = g(\bar{x}) \leq 0$. Thus $\bar{x}$ is feasible for 3.1.

For any $\in > 0$, there exists $\bar{x}$ such that $g(\bar{x}) < 0$ and $f(\bar{x}) \leq f(x^*) + \in$. For each k,

$$f(x^*) + \in + \mu_k b(\bar{x}) \geq f(\bar{x}) + \mu_k b(\bar{x}) \geq F(\mu_k, x^k)$$

Therefore for k sufficiently large, $f(x^*) + 2\in \geq F(\mu_k, x^k)$, and since $F(\mu_k, x^k) \geq f(x^*)$ from Lemma 3.1.4, then;

$$f(x^*) + 2\in \geq \lim_{k\to\infty} F(\mu_k, x^k) \geq f(x^*)$$

This implies that,

$$\lim_{k\to\infty} F(\mu_k, x^k) = f(x^*)$$

We also have,

$$f(x^*) \leq f(x^k) \leq f(x^k) + \mu_k b(x^k) = F(\mu_k, x^k) + 2\in \geq \lim_{k\to\infty} F(\mu_k, x^k) \geq f(x^*)$$

Taking the limits we obtained,

$$f(x^*) \leq f(\bar{x}) \leq f(x^*)$$

whereby $\bar{x}$ is an optimal solution of 3.1. Typical class of barrier functions are:

$$b(x) = \sum_{i=1}^{m} (g_i(x))^{-a} \quad \text{where} \quad a > 0$$

3.1 Review
· · · · · · · · · · · · ·

The details of Barrier and Penalty methods are discussed in this section as an optimization problem. In the Penalty method one can start with infeasibility and can ultimately be obtained feasible optimal solution. Meantime, The Barrier method uses a barrier so that the solution never becomes infeasible. These methods can be implemented without compromising computational cost. The penalty function algorithm is simple and easy to implement.

CHAPTER 4.
MOLECULAR DYNAMICS, PENALTY FUNCTION METHOD AND ITS PROPERTIES

4.0.1 Constrained molecular dynamics and penalty function method

Based on the theory of classical mechanics, the trajectory of molecular motion between two molecular states minimizes the total action of the motion (30). Let $x(t)$ be the configuration of the molecule at time $x = \{x_i : x_i = (x_{i,1}, x_{i,2}, x_{i,3})^T, i = 1,, n\}$, where x_i is the position vector of atom i and n the total number of atoms in the molecule. Given beginning and ending time t_0 and t_e, $x(t) \in [t_0, t_e]$ defines a trajectory connecting two molecular states $x_0 = x(t_0)$ and $x_e = x(t_e)$. Let $\mathbb{L}(x, x', t)$ be the difference of the kinetic and potential energy of the molecule at time t. The functional $\mathbb{L}$ is called the Lagrangian of the molecule. Let S be the action of the molecule in $[t_0; t_e]$. Then, S is defined as the integral of the Lagrangian in $[t_0; t_e]$, and according to the least action principle, the trajectory x minimizes the action S of the molecular motion in $[t_0; t_e]$:

$$min\left(S(x) = \int_{t_0}^{t_e} \mathbb{L}(x, x')dt \right)$$

(4.1)

Theorem 4.0.1. *Let* L *be a continuously differentiable functional. Let x be a solution of problem 4.1. Then, x satisfies the following Euler-Lagrange Equation:*

$$\frac{\partial \mathbb{L}(x, x', t)}{\partial x'} - \frac{d\left[\dfrac{\partial \mathbb{L}(x, x', t)}{\partial x}\right]}{dt} = 0 \tag{4.2}$$

Proof: Let δx be a small variation of x *and* $\delta x(t_0) = \delta x(t_e) = 0$. By the principle of variation, the necessary condition for x to be a solution of problem 4.1 is that:

$$\delta S = \int_{t_0}^{t_e} \left(\frac{\partial \mathbb{L}(x, x', t)}{\partial x} \delta x + \frac{\partial \mathbb{L}(x, x', t)}{\partial x'} \delta x' \right) dt = 0 \tag{4.3}$$

Since $\delta x' = \delta \dfrac{dx}{dt} = d\dfrac{\delta x}{dt}$, we obtain, after integrating the second term of 4.3 by parts:

$$\delta S = \int_{t_0}^{t_e} \left(\frac{\partial \mathbb{L}(x, x', t)}{\partial x} + \frac{d\dfrac{\partial \mathbb{L}(x, x', t)}{\partial x'}}{dt}' \right) \delta x\, dt = 0 \tag{4.4}$$

Since δS should be zero for all δx, the integrand of 4.4 must be zero and 4.2 follows.

Corollary 4.0.2. *Let* $\mathbb{L} = \dfrac{x'^{T} - Mx'}{2} - \varphi(x)$ *, where M is the mass matrix of a molecule and φ the potential energy. Then, a necessary condition for x to minimize an action S is that:*

$$Mx'' = -\nabla \varphi(x) \tag{4.5}$$

Proof: It follows from Theorem 4.1.1 and the facts that $\dfrac{d\left(\dfrac{\partial \mathbb{L}}{\partial x'}\right)}{dt} = Mx''$ and $\dfrac{\partial \mathbb{L}}{\partial x} = -\nabla \varphi$.

Equation 4.5 is well known as the equation of motion for a molecule of n atoms. It can be equivalently stated as:

$$m_i x_i'' = f_i(x_1,\ldots,x_n), \quad f_i = -\frac{\partial \varphi}{\partial x_i}, \quad i = 1,\ldots,n \tag{4.6}$$

where m_i and f_i are the mass and force for atom i, respectively and $M = diag[m_1, \ldots, m_n]$. Note that Theorem 4.1.1 and Corollary 4.1.2 imply that a trajectory that minimizes the molecular action between two system states necessarily satisfies the classical mechanical equation of motion. In other words, the solution of the equation of motion can be considered as an attempt for the minimization of the molecular action of motion.

Let $C = C_j : j = 1, ,m$ be a vector of functions that can be used to define the constraints on the molecule. The constrained simulation problem can then be considered as a constrained least action problem.

$$min \left(S(x) \ = \ \int_{t_0}^{t_e} \mathbb{L}(x, x', t) dt \right)$$
$$\text{subject to} \quad C(x) \ = \ 0 \tag{4.7}$$

Then, by the theory of constrained optimization, a necessary condition for a molecular trajectory x between x_0 and x_e to be a solution of problem 4.7 is that:

$$\delta S(x) + \sum_{j=1}^{m} \lambda_j \delta C_j(x) \ = \ 0 \tag{4.8}$$
$$C(x) \ = \ 0$$

Where λ is a vector of Lagrange multipliers.

Theorem 4.0.3. *Let* $\mathbb{L} = \dfrac{x'^T M x'}{2} - \varphi(x)$*, where M is the mass matrix of a molecule and φ the potential energy. Then, a necessary condition for x to minimize an action S subject to C(x) = 0 is that:*

$$M x'' \ = \ -\nabla \varphi(x) - C^*(x)^T \lambda$$
$$C(x) \ = \ 0 \tag{4.9}$$

where λ is a vector of Lagrange multipliers and $C^*(x)$ the Jacobian of C(x).

Proof: For $\mathbb{L} = \dfrac{x'^T M x'}{2} - \varphi(x)$**, condition 4.8 translates to:**

$$M x'' \ = \ -\nabla \varphi(x) - \sum_{j=1}^{m} \lambda_j \nabla C_j(x)$$
$$C(x) \ = \ 0 \tag{4.10}$$

and hence to 4.8 with $C^* = [\nabla C_1, \ldots, \nabla C_m]^T$.

For each atom, equation 4.9 can be written as:

$$m_i x''_i = -\nabla\varphi(x) - \sum_{j=1}^{m} \lambda_j C_{i,j}(x_1,\ldots,x_n) \tag{4.11}$$

$$C_j(x_1,\ldots x_n) = 0, \quad j=1,\ldots,m, \quad i=1,\ldots,n$$

where:

$$f_i = -\frac{\partial\varphi}{\partial x_i}, \quad C_{i,j} = \frac{\partial C_j}{\partial x_i} \quad j=1,\ldots,m, \quad i=1,\ldots,n \tag{4.12}$$

Note that in (4.11), the right-hand side of the first equation can be treated as a single force function (with the original force function plus a combination of the derivatives of the constraint functions), and therefore, the equation can be integrated in the same way as equation (1.10) by the Verlet algorithm, except that in every step, the Lagrange multipliers $\lambda_j, j = 1, \ldots,m$ have to be determined so that the new positions x_i, $i = 1, \ldots, n$ for the atoms satisfy the constraints $C_j(x_1, \ldots, x_n) = 0$, $j = 1, \ldots,m$.

Let f be the objective function and $C = \{C_j, j = 1, \ldots,m\}$ be a set of constraint functions. Consider a general equality constrained optimization problem:

$$\min\left(f(x_1,x_2,\ldots x_n)\right) \tag{4.13}$$

$$\text{subject to} \quad C_j(x_1,x_2,\ldots x_n) = 0, \quad j=1,\ldots,m$$

The unconstrained optimization problem with a quadratic penalty function for (4.13) can be defined as follows:

$$\min\left(f(x_1,x_2,\ldots x_n)\right) + \frac{\mu}{2}\sum_{j=1}^{m}\left|C_j(x_1,\ldots,x_n)\right|^2 \tag{4.14}$$

where μ is a parameter called the penalty parameter. In principle, the solution for problem (4.13) can be recovered by solving problem (4.14) with the parameter μ gradually increasing to 1. A so-called exact penalty function can also be defined, such as using the l_1-norm. Then, problem 4.14 becomes:

$$\min\left(f(x_1, x_2, \ldots x_n) + \frac{\mu}{2}\sum_{j=1}^{m}|C_j(x_1, x_2, \ldots x_n)| \right) \tag{4.15}$$

and the solution for problem (4.13) can be recovered by solving problem (4.15) with the parameter μ only raised to a sufficiently large value.

If the constraints are inequalities, i.e., $C_j(x_1, \ldots, x_n) \geq 0, \ j = 1, \ldots, m,$ the penalty functions in 4.14 and 4.15 can still be used in the same way as for equality constraints, only with C_j replaced by $C_{\bar{j}}$ for all j, where $C_{\bar{j}} = min(C_j ; 0)$ gives the amount of violation for constraint j. Another approach is to introduce a barrier function for each constraint. Then, the problem becomes minimizing the combination of the objective function and the barrier functions such as the following:

$$\min\left(f(x_1, x_2, \ldots x_n) - \tau \sum_{j=1}^{m} log\left(C_j(x_1, \ldots, x_n)\right) \right) \tag{4.16}$$

where $log(C_j(x_1, \ldots, x_n))$ is called the log barrier function for $C_j(x_1, \ldots, x_n)$ as the function is not defined when $C_j(x_1, \ldots, x_n) < 0$ and is infinity when $C_j(x_1, \ldots, x_n) = 0$. The parameter τ is used to control the barrier term. In principle, the solution of the original constrained optimization problem can be asymptotically approached by solving problem (4.16) as τ is gradually decrease to zero.

In this work, we will only use the formulation in (4.14) for the development of the penalty function method for constrained molecular dynamics simulation. The primary reasons are that in this work, we only consider the equality constraints, and the squared Euclidean norm used in (4.14) also provides smoother properties than the $l_1 - norm$ in (4.15) for optimization. By using the formulation in (4.14), the constrained least action problem as givens in (4.7) can be transformed to:

$$\min\left(S(x) + \frac{\mu}{2}\|C(x)\|^2 \right)$$

(4.17)

where $\|.\|$ is the Euclidean norm and $C = (C_1, \ldots, C_m)^T$. In principle, a solution for the constrained least action problem (4.7) can be obtained by solving a sequence of problems in (4.17) with μ selected from an increasing sequence of parameters $\{\mu_k\}$.

Theorem 4.0.4. *Let $\mu = \mu_k$ and $\mu_k \to \infty$ as $k \to \infty$. Let x^k be a global solution to (4.17) with $\mu = \mu_k$, and $x^k \to x^*$ as $k \to \infty$. Then, $C(x^k) \to 0$ as $x^k \to x^*$, and x^* is a global solution to the constrained least action problem (4.7).*

Proof: Let $\phi(x,\mu) = S(x) + \frac{\mu}{2}\|g(x)\|^2$ Then:

$$\phi(x,\mu) = S(x) + \frac{\mu}{2}\|g(x)\|^2$$

(4.18)

showing that the sequence of global minima $\phi(x^k, \mu_k)$ of (4.18) is non-decreasing. By using the facts that $\phi(x^k, \mu_k) \le \phi(x^k, \mu_{k+1})$ *and* $\phi(x^{k+1}, \mu_k) \le \phi(x^{k+1}, \mu_{k+1})$, we have

$$\phi(x^k,\mu_k) - \phi(x^{k+1},\mu_{k+1}) \le \phi(x^k,\mu_{k+1}) - \phi(x^{k+1},\mu_k)$$

(4.19)

and

$$(\mu_{k+1} - \mu_k)\left(\|C(x^k)\|^2 - \|C(x^{k+1})\|^2\right) \ge 0$$

(4.20)

It follows that $\{\|C(x^k)\|^2\}$ is non-increasing. Since $\phi(x^k, \mu_k) \leq \phi(x^k+1, \mu_k)$, $\{S(x^k)\}$ is also non-decreasing. Let S^* be the global minimum of (4.7). Then, $\phi(x^k, \mu_k) \leq \phi(x, \mu_k) = S^*$, $\{S(x^k)\}$, where x = global $argmin\{S(x) : g(x) = 0\}$. Then:

$$S(x^k) + \mu_k \|C(x^k)\|^2 \leq S^* \tag{4.21}$$

Since $\{S(x^k)\}$ is non-decreasing and $\{\mu^k\}$ is increasing, $C(x^k) \to 0$, and it follows that if $x^k \to x^*$, $C(x^*) = 0$ and $S(x^*) \leq S^*$. By the definition of S^*, $S(x^*) \geq S^*$, and therefore, $S(x^*) = S^*$. We now define an extended Lagrangian

$$\tilde{\mathbb{L}}(x,x',t) = \mathbb{L}(x,x',t) + \frac{\mu\|g(x)\|^2}{2(t_e - t_0)} \tag{4.22}$$

Then, problem (4.17) can be written in the following form:

$$min\left(\tilde{S}(x) = \int_{t_0}^{t_e} \tilde{\mathbb{L}}(x,x',t)dt\right) \tag{4.23}$$

By applying Theorem 4.1.1 and Corollary 4.1.2 to (4.23), we obtain the extended equation of motion as the necessary condition for any x to be a solution to problem (4.17),

$$Mx'' = -\nabla\varphi(x) - \mu C^*(x)^T C(x) \tag{4.24}$$

where C^* is the Jacobian of C. The following theorem shows that a solution to problem (4.7) that satisfies the necessary condition (4.9) for the problem can be obtained by solving the extended equation of motion (4.24) with μ increasing to 1. The solution is equivalent to the one that can be obtained by using a Lagrange multiplier type method.

Theorem 4.0.5. *Let $\mu = \mu_k$ and $\mu_k \to \infty$ as $k \to \infty$. Let x^k be a solution to problem (4.17) with $\mu = \mu_k$, and $x^k \to x^*$ as $k \to \infty$. Let C^* be the Jacobian of C and $C^*(x^*)$ be of full rank. Then, x^* satisfies the necessary condition (4.7) for x^* to be a solution to the constrained least action problem (4.6).*

Proof: Based on (4.24), for each pair of (x^k, μ_k), necessarily:

$$M\left[x^k\right]'' \;=\; -\nabla\varphi(x^k) - \mu_k C^*(x^k)^T C(x^k) \tag{4.25}$$

Let $\lambda_k = \mu_k C(x^k)$. Then:

$$M\left[x^k\right]'' \;=\; -\nabla\varphi(x^k) - C^*(x^k)^T \lambda_k \tag{4.26}$$

and:

$$\lambda_k = -\left(C^*(x^k)^T\right)^+ \left(M(x^k)'' + \nabla\varphi(x^k)\right) \to \; \left(C^*(x^*)^T\right)^+ \left(M(x^*)'' + \nabla\varphi(x^*)\right) - \lambda^* \tag{4.27}$$

where $(C^*(x^k)^T)^+$ is the pseudo-inverse of $C^*(x^k)^T$. Then, $C(x^k) = \dfrac{1}{\mu_k}\lambda_k \to \dfrac{1}{\mu_k}\lambda^* \to 0$. It follows that:

$$M(x^*)'' \;=\; -\nabla\varphi(x^*) - C^*(x^*)\lambda^* \quad \text{and} \quad C^*(x^*) = 0 \tag{4.28}$$

In the atomic form, equation (4.24) can be written as:

$$m_i x''_i = f(x_1,\ldots,x_n) + \mu \sum_{j=1}^{m} C_{i,j}(x_1,\ldots,x_n)\, C_j(x_1,\ldots,x_n), \quad f_i = -\frac{\partial \varphi}{\partial x_i}, \quad C_{i,j} = -\frac{\partial C_j}{\partial x_i} \quad (4.29)$$

$i = 1, \ldots, n$. By treating the entire right-hand side of each equation in (4.28) as a force function, we can then apply standard Verlet algorithms to obtain our numerical formulas for the solution of the equations in (4.28):

Penalty Position Verlet

$$x_i^{k+1} = 2x_i^k - x_i^{k-1} + \Delta t^2 (f_i^k + \tfrac{1}{m_i} \mu \sum_{j=1}^{m} C_{j,i}^k C_j^k) \quad i=1,\ldots,n, \quad k=1,\ldots \qquad (4.30)$$

Penalty Velocity Verlet

$$(4.31)$$

$$x_i^{k+1} = x_i^k + \Delta t v_i^k + \Delta t^2 \left(f_i^k + \frac{1}{2m_i} \mu \sum_{j=1}^{m} C_{j,i}^k C_j^k \right)$$

$$x_i^{k+1} = v_i^k + \Delta t \left(f_i^k + f_i^{k+1} + \frac{1}{2m_i} \mu \sum_{j=1}^{m} C_{j,i}^k C_j^k + C_{j,i}^{k+1} C_j^{k+1} \right)$$

$$i=1,\ldots,n, \quad k=1,\ldots$$

Note that formulas (4.30) and (4.31) do not involve solving nonlinear systems and can therefore be updated much more efficiently than Shake and Rattle. However, the parameter μ needs to be selected appropriately and required to be sufficiently large. There is also an issue that for different penalty terms, different scales may need to be used for the parameters. We discuss these issues in greater details in the specific implementations of the algorithms in the next sections.

4.1 Analysis of molecular dynamics

When carrying out molecular dynamic simulations, coordinates and velocities of the system are saved. These are then used for the analysis. Time dependent properties can be displayed graphically, where one of the axis corresponds to time and other to the quantity of interest, such as energy, RMSD, etc. Other approaches have been developed for representing the time dependance of angle rotation (dihedral). Average structures can be calculated and compared to experimental structures. Molecular dynamic simulations can help visualize and understand conformational changes at an atomic level when combined with molecular graphics programs which can be display the structural parameters of interest in a time dependent way. Some quantities that are routinely calculated from a molecular dynamics simulation.

4.1.1 Root Mean Square Deviation (RMSD)

Root Mean Square deviation has been implemented as a protocol for pairwise structural superposition, with atomic Euclidean distances between aligned residues being calculated along the pairwise alignment and the RMSD for the structural pair being calculated by summing the squares of these distances, dividing by the number of distances involved and calculating the root. This results in a single value with which to assess the quality of the structural alignment, and is limited in its pairwise nature.

Define two coordinate structure matrices,

$$X = \begin{pmatrix} x_{1,1} & x_{1,2} & x_{1,3} \\ x_{2,1} & x_{2,2} & x_{2,3} \\ . & . & . \\ . & . & . \\ x_{n,1} & x_{n,2} & x_{n,3} \end{pmatrix} \quad Y = \begin{pmatrix} y_{1,1} & y_{1,2} & y_{1,3} \\ y_{2,1} & y_{2,2} & y_{2,3} \\ . & . & . \\ . & . & . \\ y_{n,1} & y_{n,2} & y_{n,3} \end{pmatrix}$$

and,

where n is number of atoms. Define:

$$\|X - Y\| = \sqrt{\sum_{i=1}^{n} \sum_{j=1}^{3} (x_i^2 - x_i^2)^2} \tag{4.32}$$

translation can be calculated by;

$$\bar{X} = \begin{pmatrix} x_{1,1} & x_{1,2} & x_{1,3} \\ x_{2,1} & x_{2,2} & x_{2,3} \\ . & . & . \\ . & . & . \\ x_{n,1} & x_{n,2} & x_{n,3} \end{pmatrix} - \begin{pmatrix} x_{t,1} & x_{t,2} & x_{t,3} \\ x_{t,1} & x_{t,2} & x_{t,3} \\ . & . & . \\ . & . & . \\ x_{t,1} & x_{t,2} & x_{t,3} \end{pmatrix}$$

where

$$x_{t,j} = \sum_{i=1}^{n} x_{i,j}, \quad j = 1,2,3 \tag{4.33}$$

Then using rotation matrix $Q = \begin{pmatrix} q_{11} & q_{12} & q_{13} \\ q_{21} & q_{22} & q_{23} \\ q_{31} & q_{32} & q_{33} \end{pmatrix}$, we can calculate $RMSD$,

$$RMSD(X,Y) = \min_{Q} \left(\frac{\bar{X} - \bar{Y}Q}{\sqrt{n}} \right) \tag{4.34}$$

4.1.2 Velocity Autocorrelation Function (VAF)

The velocity autocorrelation function is a prime example of a time dependent correlation function, and is important because it reveals the underlying nature of the dynamical processes operating in a molecular system. It is constructed as follows. At a chosen origin in time we store all three components of the velocity vi, where

$$v_i^0 = \begin{pmatrix} v_{i_x}(t_0) \\ v_{i_y}(t_0) \\ v_{i_z}(t_0) \end{pmatrix} \tag{4.35}$$

for every atom i in the system. We can calculate the first contribution to the velocity auto-correlation function, corresponding to time zero. This is average of the scalar products $v_i . v_i$ for all atoms:

$$VAF(t_0) = \frac{1}{n}\sum_{i=1}^{n}\left(v_i(t_0).v_i(t_0)\right) \tag{4.36}$$

At the next time step in the simulation $t = t_0 + \Delta t$ and the corresponding velocity for each atom is:

$$v_i^1 = \begin{pmatrix} v_{i_x}(t_0 + \Delta t) \\ v_{i_y}(t_0 + \Delta t) \\ v_{i_z}(t_0 + \Delta t) \end{pmatrix} \tag{4.37}$$

and we can calculate the next point of the VAF as:

$$VAF(t_0 + \Delta t) = \frac{1}{n}\sum_{i=1}^{n}\left(v_i(t_0) . v_i(t_0 + \Delta t)\right) \tag{4.38}$$

We can repeat this procedure at each subsequent time step and so obtain a sequence of points in the VAF, as follows:

$$VAF(k\Delta t) = \frac{1}{n}\sum_{i=1}^{n}\left(v_i(t_0) \cdot v_i(t_0 + k\Delta t)\right) \tag{4.39}$$

$$VAF(k\Delta t) = \left\langle v_i(t_0), v_i(t_0 + k\Delta t)\right\rangle \tag{4.40}$$

Consider a single atom at time zero. At that instant the atom i will have a specific velocity v^0_i. If the atoms in the system did not interact with each other, the Newton's Laws of motion tell that the atom would retain this velocity for all time. This of course means that all our points VAF would have the same value, and if all the atoms behaved like this, the plot would be a horizontal line. It follows that a VAF plot that is almost horizontal, implies very weak forces are acting in the system.

On the other hand, if the forces are small but not negligible then we would expect both its magnitude and direction to change gradually under the influence of these weak forces. In this case we expect the scalar product of $v_i(t_0)$ with $v_i(t_0 + k\Delta t)$ to decrease on average, as the velocity is changed. In statistical mechanics it is called the velocity decorrelates with time, which is the same as saying the atom 'forgets' what its initial velocity was. In such a system, the VAF plot is a simple exponential decay, revealing the presence of weak forces slowly destroying the velocity correlation.

Strong forces are most evident in high density systems, such as solids and liquids, where atoms are packed closely together. In these circumstances the atoms tend to seek out locations where there is a near balance between repulsive forces and attractive forces, since this is where the atoms are most energetically stable. In solids these locations are extremely stable, and the atoms cannot escape easily from their positions. Their motion is therefore an oscillation the atom vibrate backwards and forwards, reversing their velocity at the end of each oscillation. If we now calculate the VAF, we will obtain a function that oscillates strongly from positive to negative values and back again. The oscillations will not be of equal magnitude however, but will decay in time, because there are still perturbation forces acting on the atoms to disrupt the perfection of their oscillatory motion. So what we see is a function resembling a damped harmonic motion.

Liquids behave similarly to solids, but now the atoms do not have fixed regular positions. A diffusive motion is present to destroy rapidly any oscillatory motion. The *VAF* therefore may perhaps show one very damped oscillation before decaying to zero. In simple terms this may be considered a collision between two atoms before they rebound from one another and diffuse away.

4.1.3 Ramachandran Plots

During the last stages of structure determination of proteins by any method for example x-ray crystallography, NMR, or homology modeling, structural biologists use a variety of tools, including Ramachandran plots, to call their attention to unrealistic conformations in their models. A Ramachandran plot plainly signals residues that need further work before the entire model can be declared chemically realistic.

The Ramachandran plot displays the psi and phi backbone conformational angles for each residue in a protein. The distance between two succession alpha carbon atoms in the backbone chain of a protein is approximately constant, as are the angles between the two bonds of such atoms. The proteins have only conformational freedom to rotate around the bonds in the backbone and in the side chain. The conformational angles show preferences for values that are expected based on simple energy considerations, and deviations from these angles may be used as indicators of potential error in crystallographic projects. Phi and psi angles are also used in the classification of some secondary structure elements such as beta turns.

In a Ramachandran plot, the core or allowed regions are the areas in the plot show the preferred regions for psi/phi angle pairs for residues in a protein. Presumably, if the determination of protein structure is reliable, most pairs will be in the favored regions of the plot and only a few will be in "disallowed" regions.

There are multiple definitions of the so-called core or allowed areas in Ramachandran plots.

The results of analysis can heavily depend on the definition used.

4.2 Review

In this chapter, the penalty function method is discussed. We presented theory of penalty function method. We have shown that the equation of motion can be integrated with constraints that satisfies necessary condition to have minimum for least action principle. Data that can used to analyze trajectories are also discussed such as velocity autocorrelation, RMSD and etc.

CHAPTER 5.
IMPLEMENTATION PROCEDURE

5.1 Introduction

This chapter introduces the penalty term method that we used in molecular dynamic simulations. The research work has been carried out in Department of Mathematics, Iowa State University. Simulations were performed on a 64 bit Alpha workstation with processor speed of 500Mhz, RAM 1GB and 64 bit Intel workstation of 3.60Mhz processor, RAM 1.5GB. Initial research has tested on Argon molecular system and equation of motion was simulated with Lennard-Jones potential. The details description of results of the model is discussed here. Then, the structure of Chemistry at Harvard Macromolecular Mechanics (CHARMM) program and penalty method implementation for all atom simulations are discussed.

Molecular dynamics are popular and used to calculate dynamic and equilibrium properties of complex protein system or cluster of atoms that might not able to estimate analytically. It represents interface between experiments and theory of trajectory or motion of the system with classical mechanics and statistics theory. To obtain more complete understanding of protein, it is essential to have detailed knowledge of their dynamics. The motivation for using classical mechanics with penalty function method is the dreadful exponential scaling of the computational resources needed (CPU time and memory) with the size of the system. Yet it can be shown that for many thermodynamic systems at reasonable temperatures classical mechanics make a fairly good approximation. The penalty function method is an optimization method that we used to find minimum/maximum of the system by converting constraints optimization problem into sequence of unconstraint optimization problems. The method of penalty functions is simple and effective, provided that suitable values for the parameters can be chosen and some numerical trial and error is often necessary. One of the main advantage of this method is that simulation can be started with infeasible solution set. Most practical applications have an infeasible starting point. The dynamics are carried out with the penalty function method as an initial value problem since it satisfies necessary conditions for minimization problem. These types of problems are called least action problems.

In next section, We present penalty function method and it's implementation on Argon clusters. Distance constraints are used. Results are shown that we can increase size of the time step by introducing constraints. It also spend significantly less computing time in dynamic simulations compare to other typical dynamic simulation methods to reach equilibrium.

5.1.1 Penalty function implementation on Argon clusters

As described in chapter 1, the Van der Waals potential characterizes the contribution of the non-bonded pairwise interactions between atoms. It is generally described by the Lennard-Jones potential function. The Lennard-Jones potential is a key part of many empirical energy models, including all commonly used energy functions for proteins. A system containing more than one atom, whose Van der Waals interaction can be described by Lennard-Jones potentialis called a Lennard-Jones cluster:

$$\varphi(x) = 4\in \sum_{i<j}\left(\frac{\sigma^{12}}{r_{i,j}^{12}} - \frac{\sigma^{6}}{r_{i,j}^{6}}\right) \tag{5.1}$$

where $\sigma = 0.405A$ and $\in = 165.4e^{-23}$ J. The Lennard-Jones potential function for a single pair of neutral atoms is a simple uni-modal function. This is illustrated by Figure (1.7). It is easy to find the overall minimum of this function that is assumed at 1 with energy -1. In a complex system, many atoms interact and we need to sum up the Lennard Jones potential functions for each pair of atoms in a cluster. The result is a complex energy landscape with many local minima. The Lennard Jones potential can be written as:

$$\varphi(x) = 4\in \sum_{1\leq i\leq j\leq 3}\left(\frac{\sigma^{12}}{r_{i,j}^{12}} - \frac{\sigma^{6}}{r_{i,j}^{6}}\right) \tag{5.2}$$

If one uses $i \neq j$, the total energy must be divided by two. The Lennard Jones potential function is partially separable (A function that is the sum of functions, each of which only involves a disjoint subset of the variables, is called partially separable.). The partially separability of the Lennard Jones potential implies that, if a single atom or molecule in a cluster is moved, the potential energy can be re-evaluated cheaply at a cost that is only $\left(\frac{2}{n}\right)^{th}$ of the cost of a total function evaluation, where n is the total number of atoms or molecules in the cluster. This is due to the fact that the potential function composed as the sum of pairwise interactions between atoms or molecules. Given a cluster of n atoms, the Lennard Jones cluster problem is to find the relative position of atoms in the three-dimensional Euclidean space that represent a potential energy minimum.

Let $x_i = (x_{i1}, x_{i2}, x_{i3})^T$ represent the coordinates of atom i in the three-dimensional Euclidean space. Let $S = ((x_1)^T, ..., (x_n)^T)^T$, where n is the number of atoms in the cluster. The Lennard Jones potential of a pair of atoms (i, j) is:

$$\varphi(x_{i,j}) = 4 \in \left(\frac{\sigma^{12}}{r_{i,j}^{12}} - \frac{\sigma^6}{r_{i,j}^6} \right) \tag{5.3}$$

Where $r_{i,j} = \|x_i - x_j\|$. The Lennard Jones cluster problem described in the previous section can be formulated in the coordinate space as follows:

$$\varphi(S) = \sum_{i<j} \varphi\left(\|x_i - x_j\|\right) \tag{5.4}$$

$$= 4 \in \sum_{i=1}^{n-1} \sum_{j=i+1}^{n} \left(\frac{\sigma^{12}}{\|x_i - x_j\|^{12}} - \frac{\sigma^6}{\|x_i - x_j\|^6} \right) \tag{5.5}$$

where x_i and x_j represent the coordinates of the ith and the jth atoms, respectively. As it is illustrated by Figure (1.7), for a single pair of neutral atoms, the overall potential energy minimum is reached when the distance between two atoms is one. When this distance approaches zero, the potential tends to infinity. When an atom is far away from the system, its contribution to the total potential becomes almost zero. Due to these observations, it is reasonable to expect that at the optimal solution of the Lennard Jones cluster problem all atoms in $\mathbb{R}^3$ are close to unit distance to each other. However, complexity of determining the global minimum energy of the Lennard Jones cluster belongs to the class of NP-hard problem (51). In other words, there is no known algorithm that can solve this problem in polynomial time. The main difficulty in solving the Lennard Jones minimization problem arises from the fact that the objective function is a non-convex function of many variables with a large number of local minima. This non-convexity makes it very difficult to find global optimal solutions. The potential function in (5.5) can be used to describe Argon molecule cluster with equation of motion since Argon molecules have only non-bond interactions.

$$m_i \ddot{x}_{i,j} = -\nabla \varphi(S)$$

(5.6)

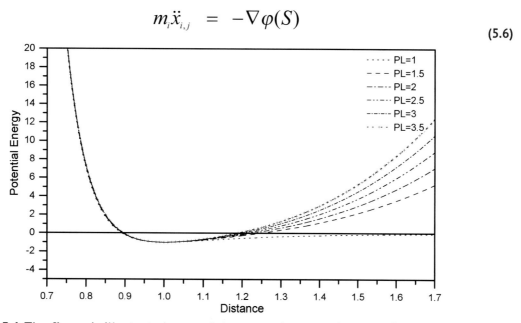

Figure 5.1 The figure is illustrated potential energy changes when penalty term change.

The Argon molecules have only non-bond interactions. Therefore, in implementation, the bond lengths are used as a constraints. Define:

$$C = \sum_{\text{some } i} \left\| r_{i,j}^2 - d_{i,j}^2 \right\|^2$$

(5.7)

where $r_{i,j}$ is distance between i^{th} and j^{th} atoms in $\mathbb{R}^3$ and $d_{i,j}$ is the target (optimal) distance between i^{th} and j^{th} atoms. The number of constrained included in simulation need to determined in the beginning. If one choose all the constraints then, the system is more rigid while less constrained allowed exibility of the system. The figure (5.2) shows iterative procedure of algorithm. The constraints optimization problem could be defined as:

$$min\left(\varphi(x_{i,j})\right)$$

such that $\quad C = 0$

(5.8)

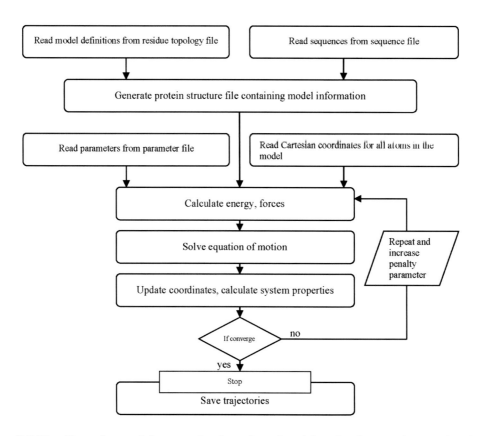

Figure 5.2 The flow chart of the penalty function algorithm for Argon cluster simulation.

Then, the constrained optimization problem can be converted into unconstrained optimization by:

$$F(x_{i,j}) \;=\; \varphi(r_{i,j}) + \mu C(x) \tag{5.9}$$

$$F(x_{i,j}) \;=\; \left(\frac{1}{r_{i,j}^{12}} - \frac{2}{r_{i,j}^{6}} \right) + \mu \sum_{\text{some } i} \left\| r_{i,j}^{2} - d_{i,j}^{2} \right\|^{2} \tag{5.10}$$

where μ is Penalty parameter. The negative gradient of equation (5.10) is used as a force in the equation of motion. In figure (5.1) generated by assuming that there are only two atoms in the system and have constraints distance between them is 1. It shows how potential energy changes with different (increase) penalty term.

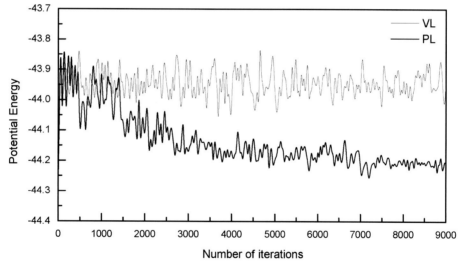

Figure 5.3 Changes in potential energy of the trajectory for argon cluster 13 produced by the penalty function method. Here, randomly selected 60% of all distances were constrained to their distances in the global energy minimum configuration. The trajectory already approached to the global energy minimum (-44.3) of the cluster in 3000 time steps while the trajectory generated by the Verlet remained in high energy. The time step is 0.032*ps* and penalty term updated every 500 iteration by 1.

The algorithm has been developed in high performance Fortran 90 (Appendix B). Simulations are performed in high performance computer with 48 processors. A serial code is used for verification (Appendix A) purpose. The Message Passing Interface (MPI) used for communications between nodes. The simulations are focused on trajectory around the global minimum of Argon atom clusters. The initial structure and velocity of clusters are generated by perturbing the global minimum structure and using Gaussian distribution function respectively. The algorithm is developed in such a way that it can use all the bond-length constraints or part of them.

Each processor is asked to perform an independent simulation with different initial structure and velocities. The penalty parameter is increased gradually once the simulation is in progress. After every iteration, we investigate potential energy changes with the previous step. If there is no improvement in the potential energy, even after increasing the penalty term, then the program is terminated. Computing time mainly depends on number of atoms in the cluster if uses same time step. The simulations were performed on most of the structures where global minimum was known (35). The simulation procedure is best described in figure (5.2). The selected number of atom cluster simulation results are presented in this section, specially 13, 24 Argon atoms.

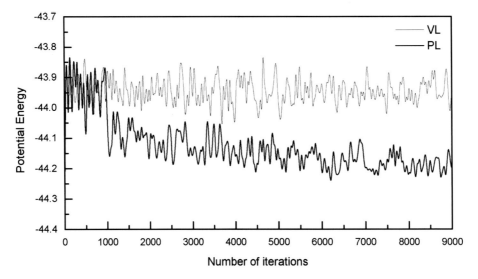

Figure 5.4 Changes in potential energy of the trajectory for argon cluster 13 produced by the penalty function method. Randomly selected 60% of all distances were constrained to their distances in the global energy minimum configuration. The time step is 0.032*ps* and penalty term updated every 1000 iteration by 5.

In figure (5.3) and (5.4) shows potential energy changes when simulation proceed with increase of penalty term. In both simulations 13 Argon atoms are selected with common time step $\Delta t = 0.032ps$. 60% of bond length constrained are selected. Even though dynamic simulations are carried out for longer time, the 9000 (9000 × 0.032ps) iterations results are presented. The simulation describe in figure (5.3) - simulation A - changes penalty term by 1 in every 500 iterations while figure (5.4) - simulation B - simulation changes penalty term by 5 in every 1000 iterations. The both A and B are shown rapid decrease of potential energy over time. But A run, the potential energy drop gradually compare to simulation B. There is small but significant variation of potential energy in simulation A. During testing, we recognized that system need to run for a sufficient time between increase of penalty term. This time is enable energy to convert kinetic to potential and vice versa.

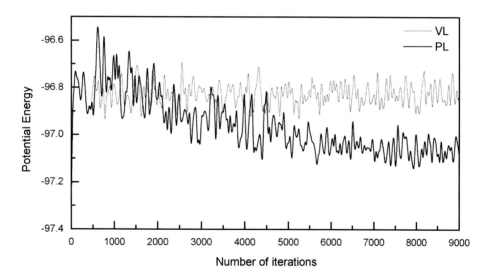

Figure 5.5 Changes in potential energy of argon cluster 24. Solid and dotted lines show the potential energy of the trajectory produced by the Verlet (VL) and penalty function (PL) methods, respectively. Here, randomly selected 50% of all distances were constrained to their distances in the global energy minimum configuration (-97.349).

In figure (5.5), potential energy of Verlet run and Penalty run are plotted for a system with 24 Argon atoms. They have the same starting structure and initial velocities. The bold and light lines are represented Penalty and Verlet runs respectively. The Penalty term is increased in every 500 iterations by 1. Freezing bond length constraints, the Argon molecules approximately reach it known global potential energy level while Verlet does not reach lower energy configuration even for long enough simulation.

We implemented penalty function method on popular molecular dynamic simulation called CHARMM and tested for BPTI (4PTI) protein. The detailed analysis of the simulation is discussed. The Verlet and Shake schemes are performed parallel to Penalty scheme for comparison purposes.

5.2 CHARMM settings

Chemistry at Harvard Macromolecular Mechanics (CHARMM) is a highly exible molecular mechanics and dynamics program originally developed by Dr. Martin Karplus at Harvard University (9). A variety of systems, from isolated small molecules to solvated complexes of large biological macromolecules, can be simulated using CHARMM. It uses empirical energy functions to describe the forces on atoms in molecules. These functions, plus the parameters for the functions, constitute the CHARMM force field. Well-validated energy and force calculations form the core of a broad range of calculation and simulation capabilities, including calculation of interaction and conformational energies, local minima, barriers to rotation, time-dependent dynamic behavior, free energy, and vibrational frequencies. The CHARMM process including penalty function can be summarized in Figure (5.6). The steps can be described in following ways:

Read model definitions: Information about residues, the basic chemical units that comprise all models, is stored in residue topology files (.RTF). The atoms, atomic properties, bonds, bond angles, torsion angles, improper torsion angles, hydrogen bond donors, acceptors, and antecedents, and non-bonded exclusions are all specified on a per residue basis.

Read sequence: Sequence information must be supplied from sequence (.seq) files or include in input file before a model can be simulated.

Read parameters: After a structure has been generated, its energy can be evaluated only if parameters exist for all internal, external, and special energy terms. Parameter files contain parameters that specify force constants, equilibrium geometries, Van derWaals radii, and other data needed for calculating energies. The values are derived from experimental data and quantum mechanical calculations. Refinement and extension of parameters are continuing process.

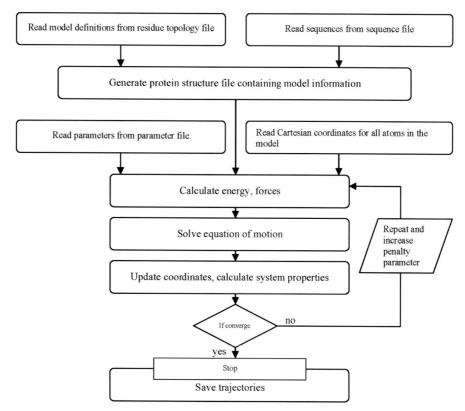

Figure 5.6 CHARMM simulation procedure.

Generate .PSF file: The protein structure file (.PSF) is the concatenation of information in the .RTF file. It specifies the information for the entire structure. The .PSF file has a hierarchical organization with atoms collected into groups, groups into residues, and residues into segments that comprise the structure. Each atom is uniquely identified within a residue by its IUPAC name, residue identifier, and segment identifier.

Read or generate Cartesian coordinates: Cartesian coordinates can be read into the coordinate file or generated from internal coordinates and parameter files. Internal coordinate files contain information about the relative positions of atoms within a structure. Two sets of Cartesian coordinates are provided. The main set is the default used for all operations involving the positions of atoms. A comparison or reference set is used for a variety of purposes, such as a reference for rotation or for operations that involve differences between coordinates for a particular molecule. Associated with each coordinate is a general purpose weighting array.

Calculate energy: The main purpose of CHARMM is the evaluation and manipulation of potential energy of a macromolecular system. Before the energy of a structure can be evaluated and manipulated, the .PSF file for the structure generated from the appropriate. RTF file, All parameters required by the .PSF file and Cartesian coordinates for every atom in the structure must be available.

Iteratively perform calculations and simulations: Using information in the .PSF, parameters file, and the energy data, any of a number of things can be done at this point including molecular dynamics, free energy perturbation, and imposing periodic boundaries. If convergence criteria is not satisfied then repeat the procedure while increasing penalty term.

A typical molecular dynamics run involves six basic steps (figure (5.7)) described as followed:

Preliminary preparation: A molecular structure with all Cartesian coordinates defined is required for a dynamics simulation. After determining the internal coordinate values of the molecule, total energy as a function of the Cartesian coordinates is computed by evaluating the individual terms of the energy equation.

Minimization: All dynamics simulations begin with an initial structure that may be derived from experimental data. Energy minimization is performed on structures prior to dynamics to relax the conformation and remove steric overlap that can produce bad contacts. In the absence of an experimental structure, a minimized ideal geometry can be used as a starting point.

Heating: A minimized structure represents the molecule at a temperature close to absolute zero. Heating is accomplished by initially assigning random velocities according to a Gaussian distribution appropriate low temperature and then running dynamics. The temperature is gradually increased by assigning greater random velocities to each atom at predetermined time intervals.

Equilibration: Equilibration is achieved by allowing the system to evolve spontaneously for a period of time and integrating the equations of motion until the average temperature and structure remain stable. This is facilitated by periodically reassigning velocities appropriate to the desired temperature. Generally, the procedure is continued until various statistical properties of the system become independent of time.

Production: In the final molecular dynamics simulation, CHARMM takes the equilibrated structure as its starting point. In a typical simulation, the trajectory traces the motions of the molecule through a period of at least 10 picoseconds. Just as with energy minimization, provision is made to update the non-bonded and hydrogen bonded lists periodically. Additional options are available, making the dynamics facility quite flexible.

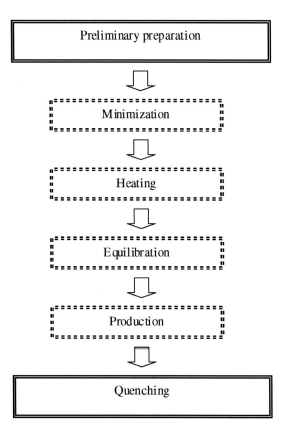

Figure 5.7 Basic steps of molecular dynamic simulation procedure.

Quenching: The logical opposite of heating, this optional step takes the molecule from the equilibrated temperature to zero. Quenching is a form of minimization, utilizing molecular dynamics to slowly remove all kinetic energy from the system.

Sometimes, minimization and heating are not necessary, provided the equilibration process is long enough. However, these steps can serve as a means to arrive at an equilibrated structure in an effective way. A molecular dynamics run generates a dynamics trajectory consisting of a set of frames of coordinates and velocities that represent the trajectory of the atoms over time. Using trajectory data, we can compute the average structure and analyze fluctuations of geometric parameters, thermodynamics properties, and time-dependent processes of the molecule. Preliminary analysis is possible using commands provided in the coordinate manipulation facility. Gross changes, as well as more detailed perturbations, can be monitored using correlation functions. Because molecular dynamics runs often require considerable amounts of computer time, a restart facility is available that allows to suspend the simulation and resume the calculation.

5.2.1 CHARMM minimization energy process

The goal of energy minimization is to find a set of coordinates representing a molecular conformation such that the potential energy of the system is at a minimum. As a consequence of many degrees of freedom for even the simplest of macromolecules, this task can be computationally quite difficult. CHARMM (9) has five different minimization methods.

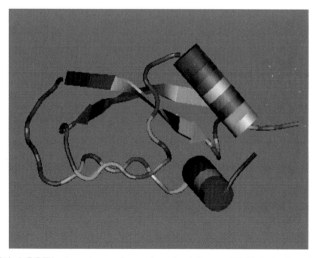

Figure 5.8 Initial BPTI structure downloaded from PDB data bank. Picture uses display style cartoon, coloring is based on RESID and use VMD software.

These four methods are provided a flexible array of iterative methods to facilitate energy minimization. Although the resulting conformation may only represent a local minimum, even macromolecules can be energy minimized efficiently using a number of these techniques. All of the minimization methods take a molecular structure to a local minimum in the potential energy surface. There is no guarantee that this will be a global minimum. Small molecular systems can be minimized to a global minimum, but multiple runs from different starting points should be done to confirm that a global minimum has indeed been found. With macromolecules, a very low probability exists that a local minimum will be the global minimum. In fact, a global minimum may never be found because of the complexity of the potential energy surface. Minimization is an important tool in analyzing proteins that are generated through site-directed mutagenesis. After substituting, inserting, or deleting residues in a sequence, minimization, along with side-chain conformation scanning, can be used to determine whether the resulting mutant structure is very much perturbed with respect to the wild type. If the perturbation is minimal, it is possible to model the structure of the mutant protein without resorting to X-ray diffraction studies.

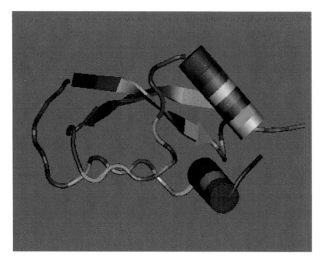

Figure 5.9 BPTI with four water molecules. Picture uses display style CPK. VMD software is used to create picture. Color is based on RESID.

5.2.2 Minimization methods

Each of the minimization methods available in CHARMM, together with implementation considerations are listed below:

1. **Steepest Descents:**

This is a very simple method. Uses only first derivative information and saves only the current location of the coordinates from iteration to iteration. In general, steepest descents converges very slowly to a local minimum in a complex potential energy surface. This method is very useful for small changes, such as the removal of unfavorable steric contacts.

2. **Conjugate Gradient:**

Exhibits better convergence than the steepest descents method. It is iterative and makes use of the previous history of minimization steps and the current gradient to determine the next step.

3. **Powell:**

A variation of the conjugate gradient method with improved efficiency. This is useful whenever the Adopted Basis-set Newton-Raphson method (described below) is not possible.

4. **Newton-Raphson:**

Implementation in CHARMM involves diagonalization of the second derivative matrix, then finding the optimal step size along each eigenvector. When one or more negative eigenvalues exist, a blind application of the equations will find a saddle point in the potential. To overcome this problem, a single additional energy and gradient determination is performed along the eigenvector displacement for each small or negative eigenvalue.

From this additional data, the energy function is approximated by a cubic potential and the step size that minimizes this function is adopted. The advantages of this algorithm are that it avoids saddle points in the potential energy surface and converges rapidly when the potential is nearly quadratic. The major disadvantage is that large computational requirements makes this technique time consuming and memory demanding for large molecules.

5. **Adopted Basis-Set Newton-Raphson:**

Similar to conjugate gradients, but fewer energy evaluations are usually necessary because the linear interpolation phase of conjugate gradients is avoided. This method performs energy minimization using a Newton-Raphson algorithm applied to a subspace of the coordinate vector spanned by the displacement coordinates of the last positions. The second derivative matrix is constructed numerically from the change in the gradient vectors, and is inverted by an eigenvector analysis that allows the routine to recognize and avoid saddle points in the energy surface. At each step, the residual gradient vector is calculated and used to add a steepest descent step, incorporating new direction into the basis set. This method is the method of choice for most applications. Because it avoids the large storage requirements.

6. **Truncated-Newton (TN) Minimization Package:**

This method was developed by T. Schlick and A. Fogelson. TNPACK is based on the preconditioned linear conjugate-gradient technique for solving the Newton equations. The structure of the problem (sparsity of the Hessian) is exploited for preconditioning. TNPACK can converge more rapidly than ABNR for small and medium systems (up to 400 atoms) as well as large molecules that have reasonably good starting conformations.

5.2.3 CHARMM force field

The CHARMM potential energy function is defined as follows;

$$\varphi = \sum_{bonds} k_b(b-b_0)^2 + \sum_{angles} k_\theta(\theta-\theta_0)^2 + \sum_{Dihedrals} k_\phi\left(1+\cos(n\phi-\delta)\right) + \qquad (5.11)$$

$$\sum_{impropers} k_\omega(\omega-\omega_0)^2 + \sum_{Urey-Bradley} k_u(u-u_0)^2 +$$

$$\sum_{Non-bonded} \in_{i,j}\left(\left(\frac{R^{min}}{r_{i,j}}\right)^{12} - \left(\frac{R^{min}}{r_{i,j}}\right)^{6}\right) + \frac{q_i q_j}{\in r_{i,j}}$$

There are several versions of the CHARMM force field. We used CHARMM22 (released in 1991). The first term in the energy function accounts for the bond stretches where k_b is the bond force constant and $(b - b_0)$ is the distance from equilibrium that the atoms have moved. The second term in the equation accounts for the bond angles where k_θ is the angle force constant and $(\theta - \theta_0)$ is the angle from equilibrium between three bonded atoms. The third term is for the dihedrals where k_ϕ is the dihedral force constant and n is the multiplicity of the function, is the dihedral angle and is the phase shift. The fourth term accounts for the improper angles, that are out of plane bending, where k_ω is the force constant and $(\omega - \omega_0)$ is the out of plane angle. The Urey-Bradley component comprises the fifth term, where k_u is the respective force constant and u is the distance between the first and third atoms in the harmonic potential. Non-bonded interactions between (i, j) pairs of atoms are represented by the last two terms. By definition, the non-bonded forces are only applied to atom pairs separated by at least three bonds. The van Der Waals energy is calculated with a standard 12-6 Lennard-Jones potential and the electrostatic energy with a Coulomb potential. In the Lennard-Jones potential above, the R^{min} term is not the minimum of the potential, but rather where the Lennard-Jones potential crosses the x-axis.

5.2.4 Convergence criteria

As minimization is proceeding, CHARMM computes the values of several terms that can be monitored for energy convergence. These are:

- Root mean square (RMS) gradient
- Step size
- Energy change

If any of these terms is smaller than the default or the user-defined tolerance, minimization will stop. Although a zero RMS gradient is a necessary condition for a minimum, it is not a satisfying condition.

All energy minimizations are involved calculating the potential energy of the system. One must have a .PSF, coordinates, and a parameter file available prior to minimization. Hydrogen bonded and non-bonded lists must also be created prior to any energy evaluation and subsequent minimization.

5.3 Penalty method implementation

The CHARMM (9) program is modified to implement Penalty function method. The three different molecular dynamic simulations have been performed. One with Verlet (VL) scheme, other two with Shake (SH) scheme and Penalty (PL) scheme and they all use bond length as a constraints. There are no external solvent molecules are included. The bovine pancreatic trypsin inhibitor (BPTI) (Figure (1.3)) is selected to investigate efficiency of those methods. This molecule was chosen for study because in literature there have been number of previous simulations of its dynamic properties [(32), (23), (26), (31)].

Title: **The Geometry of the Reactive Site and of the Peptide Groups in Trypsin, Trypsinogen and its Complexes with Inhibitors**
Compound: **Trypsin Inhibitor**
Authors: **R. Huber, D. Kukla, A. Ruehlmann, O. Epp, H. Formanek, J. Deisenhofer, W. Steigemann**
Exp. Method: **X-ray Diffraction**
Classification: **Proteinase Inhibitor (Trypsin)**
Source: **Bos taurus**
Common name:**domestic cattle, domestic cow, cattle**
Deposition Date: **27-Sep-1982**
Release Date: **18-Jan-1983**
Resolution [Å]: **1.50**
R-Value: **0.162**
Residues: **58**
Atoms: **514 (454 + water molecules)**
Sequence:
ARG PRO ASP PHE CYS LEU GLU PRO PRO TYR THR GLY PRO CYS LYS ALA ARG ILE ILE ARG
TYR PHE TYR ASN ALA LYS ALA GLY LEU CYS GLN THR PHE VAL TYR GLY GLY CYS ARG
ALA LYS ARG ASN ASN PHE LYS SER ALA GLU ASP CYS MET ARG THR CYS GLY GLY ALA

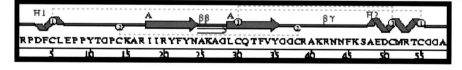

Figure 5.10 The figure is showed sequence of BPTI (9).

To compare the three molecular dynamic simulations and determine whether or not they sample approximately the same part of phase space, a verity of statistical properties are analyzed. They included the averages, fluctuations and correlation functions for various physical quantities.

Following units are used in this book:

Time: Pico seconds (ps) [1 $ps = 10^{-12}$ *seconds]*

Temperature: Kelvin (K)

Mass: Atomic mass units (u)

Length: Angstrom (Å)

Energy: Kilocalorie per molecules (*kcal mol^{-1}*)

We used (equation 5.12) in the implementation of the penalty function method in CHARMM.

Figure 5.11 This picture shows BPTI with all hydrogen atoms. There are 904 atoms in total. Picture uses display style CPK and coloring is based on RESID.

The penalized energy function becomes the following:

$$\varphi = \mu \sum_{bonds} k_b (b-b_0)^2 + \sum_{angles} k_\theta (\theta - \theta_0)^2 + \sum_{Dihedrals} k_\phi \left(1 + \cos(n\phi - \delta)\right) + \qquad (5.12)$$

$$+ \sum_{impropers} k_\omega (\omega - \omega_0)^2 + \sum_{Urey-Bradley} k_u (u - u_0)^2 +$$

$$\sum_{Non-bonded} \in_{i,j} \left(\left(\frac{R^{min}}{r_{i,j}} \right)^{12} - \left(\frac{R^{min}}{r_{i,j}} \right)^6 \right) + \frac{q_i q_j}{\in r_{i,j}}$$

where the original bond-length energy (the first term) is replaced by a penalty function for the bond length constraints. Note that the penalty term for each bond-length is multiplied by a constant $k_{i,j}$. The term can then be scaled by using an appropriate value for $k_{i,j}$. In our implementation, we simply used the corresponding force constant for each $k_{i,j}$. Coincidentally, the penalized energy function then becomes exactly the original energy function when $\mu = 1$ and is a continuation from the original energy function for any $\mu > 1$. In our implementation, the penalty parameter was changed gradually from value (0.7) less than 1 to a value (1.7) beyond 1 during the simulation.

Protein BPTI (figure (1.3)) is contained 58 amino acid residues. It consists of 454 atoms. In addition, four internally hydrogen bonded water molecule are included in the simulations, making total number of atoms equal to 458 (without hydrogen) (7). When bond-length constraints are applied, the bond stretching potential term is omitted and all bond lengths except hydrogen bonds of the protein are kept fixed. The VL, SH and PL runs, an integrating time step $\Delta t = 10^{-3} ps$ have been chosen. Moreover, the $\Delta t = 2 \times 10^{-3} ps$ and other larger time steps are used in VL run (23). In SH and PL run the relative accuracy tolerance to which the constraints are to satisfied geometrically must be specified. However, dynamical accuracy of SH and PL depend not only tolerance but also Δt. SH runs, the tolerance has been chosen as small as 10^{-5}.

The initial BPTI protein system obtained from X-ray structure. The data is downloaded from Protein Data Bank (7), PDB - http://www.rcsb.org/pdb/, figure (5.10)) which contained 454 atoms and 60 water molecules (_gure (5.8)). Out of 60 water molecules, carefully selected internal four molecules added to protein. This has been done with program called **gOpenMol** (http://www.csc.fi/gopenmol/). Then, hydrogen bonds are added to the system and build a three dimension structure using CHARMM.

The potential energy of the the system minimized by applying steepest descent method. Before minimize BPTI has 44906.75 *kcal mol^{-1}* of potential energy. The energy is minimized until decrease less than 10^{-3} *kcalmol^{-1}*. This occurred after 2999 steps and spent elapsed time 11.97 minutes and cpu time 3.28 minutes on Alpha 500Mhz 64 bit processor.

Cycle	Energy	Step-size
2998	-1137.46888	0.00034
2999	-1137.46900	0.00041

Table 5.1 Final steps of energy minimization

In 2999 step, the time step is less than 1×10^{-3} and total energy is -1137.49 *kcal mol^{-1}* (table (5.1)). This part is carried out to eliminate the strain present in X-ray structure.

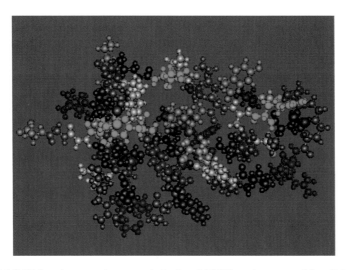

Figure 5.12 This picture shows minimized BPTI stricture with all hydrogen atoms. Display style is CPK and coloring is based on RESID. VMD is used.

Heating was accomplished by initially assigning random velocities to atoms according to a Gaussian distribution appropriate for that low temperature and then running dynamics simulation with VL. The temperature was then increased gradually by assigning greater random velocities to atoms at every $0.05ps$ from absolute zero ($3.42K$) to $300K$. The entire heating process used 5000 simulation steps with $0:001ps$ time step, which is corresponded to total $5ps$ simulation time (figure 5.14). When simulation started, the temperature rose rapidly. The conversion of kinetic energy to potential energy was fast. However, the increase in temperature decreased when the system aged.

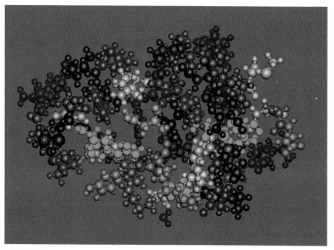

Figure 5.13 Average of $25ps$ structure of equilibrium period of BPTI structure including all hydrogen atoms. CPK display style and color is based on RESID. The picture is created by using VMD software.

To achieve the equilibrium state for SH and PL, we first performed $15ps$ and $20ps$ simulations with VL and then started SH and PL with initial positions and velocities taken from the final step of VL respectively (figure 5.14). We then ran SH and PL for $25ps$ for analysis (figure 5.14). The computing time for each simulation is presented in Table 1. VL, SH and PL are required 2:44, 3:00 and 2:44 minutes of computing time per picosecond simulation on an Alpha workstation. We recorded the coordinates of the trajectories every $0:01ps$. The results in the final $25ps$ of the simulations were used to calculate dynamical and statistical properties of the system.

The bond length constraints are read from .PARM _le in CHARMM. In .PARM file, the standard optimal distances are defined for each types of molecular bonds.

Average root mean square deviation of 25*ps* three simulations (VL, SH and PL) of Backboneatoms is shown in table (5.3. Two constraints methods SH and PL are showed lowest RMSD while PL and VL has lowest RMSD compare to SH and VL.

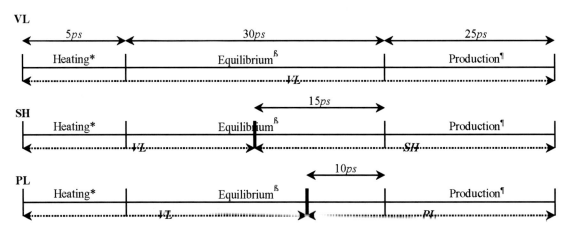

Figure 5.14 Simulation time for VL, SH and PL. *Heating - bring the system to normal temperature; §Equilibrium - the time for the system to reach the equilibrium; ¶Production - stable dynamic results for analysis.

Scheme	*Computing time
VL	1.14 hours
SH	1.25 hours
PL	1.14 hours

Table 5.2 Computing time of VL, SH and PL run. *Computing time for the 25ps simulation after equilibrium

All the averages and correlation functions are presented in the next chapter are from final 25*ps* of simulations period. The coordinates of trajectory are saved every 0.01*ps* and carefully studied.

	X-ray	VL	SH	PL
X-ray	0	1.786	1.726	1.673
VL		0	1.156	0.962
SH			0	0.611
PL				0

Table 5.3 Root mean square deviation (RMSD) of backbone atoms

5.4 Review

In this chapter, implementation of penalty function method is discussed. In addition, the standard molecular dynamic simulation procedures are presented. A test case of Argon is presented. Part of CHARMM program is highlighted and Argon and BPTI simulation results are discussed.

CHAPTER 6.
RESULTS, SUMMARY AND DISCUSSION

6.0.1 Analysis of dynamics

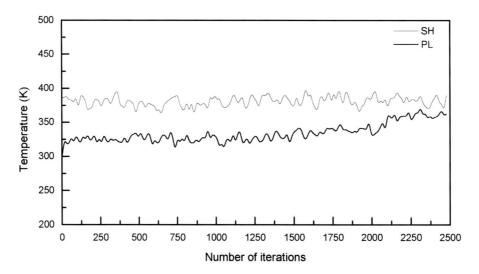

Figure 6.1 Temperature distribution of Shake and Penalty run.

Figure (6.1) shows the temperature distribution in the 25ps simulation by SH and PL. The variation of the temperature showed that there was a difference between SH and PL at the beginning of the simulation. The temperature for PL started at 300K, the same as VL, but gradually increased and eventually approached to that for SH. This indicated that the simulation by PL started with a condition similar to that by VL but then changed to SH later when the penalty parameter is fully adjusted to an appropriate value. Moreover, figure (6.2) shows temperature distribution of VL run for 25*ps*. Fluctuation of temperature is in range of 25*K*.

The average backbone root mean square (RMS) fluctuations are plotted as a function of residue number in _gure (6.3). The graphs show a great correlation between the fluctuations by SH and PL. On the other hand, VL simulation produced large fluctuations for 10 TYR, 13 PRO, 15 LYS, 27 ALA, 45 PHE and 47 SER residues, which were disagreed with those by SH and PL.

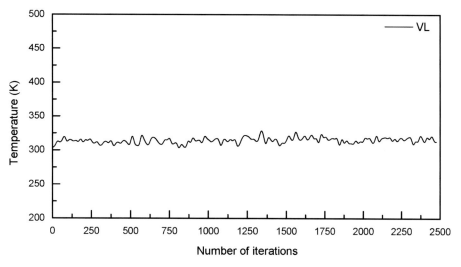

Figure 6.2 Temperature distribution of Verlet run.

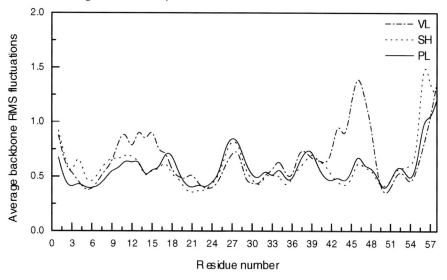

Figure 6.3 The average backbone RMS fluctuations of the residues in the *25ps* production simulations.

The root mean square fluctuations of C_α atoms in the simulations are plotted in figure (6.4). Similar to the average backbone root mean square fluctuations, the C_α fluctuations by PL and SH again had strong correlations.

The average root mean square fluctuations of *HN* and the non-backbone atoms by SH and PL correlated as well shown in figure (6.5) and (6.6) except for some discrepancies around residues 54 to 58.

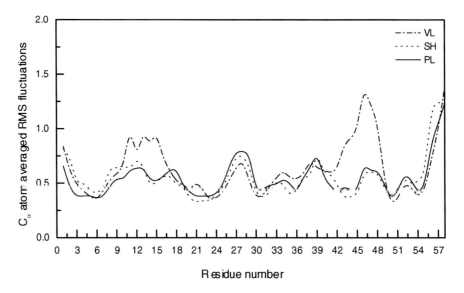

Figure 6.4 The average C_α RMS fluctuations in the *25ps* production simulations.

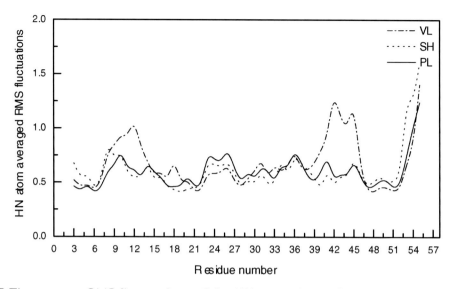

Figure 6.5 The average RMS fluctuations of the HN atoms in the 25*ps* production simulations.

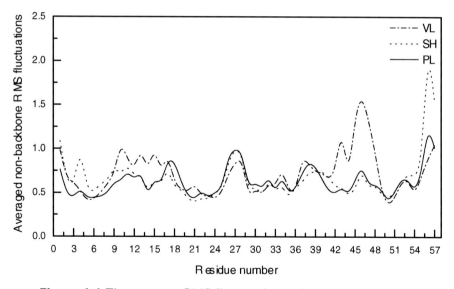

Figure 6.6 The average RMS fluctuations of the non-backbone
atoms in the 25*ps* production simulations.

Figure (6.7) shows the normalized velocity autocorrelations calculated for 51 CYS using the trajectories produced by VL, SH, and PL. For demonstration purposes, the correlations over a 10ps time period are shown. The first curve is for VL run with an autocorrelation time equal to 0:01*ps*. The auto correlation time for the second curve is 0:02*ps* and is half the resolution of the first one. The third and fourth curves are for SH and PL runs, respectively, both with the autocorrelation time equal to 0:01*ps*. The curves for SH and PL showed similar correlations with that for VL in 0:02*ps* resolution, suggesting that both SH and PL are roughly faster than VL by two folds.

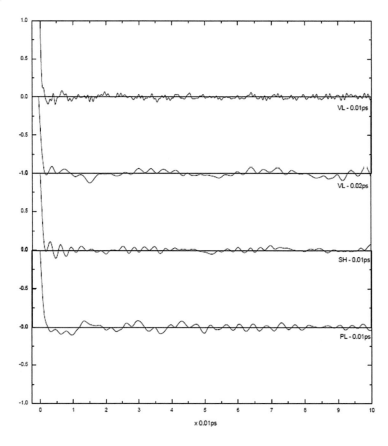

Figure 6.7 The velocity auto correlations of the C_α atom of 51 *CY S* based on the trajectories produced by VL, SH, and PL in a time period of 0.1*ps*.

6.1 Review
· · · · · · · · · · · ·

We compare VL SH and PL to determine the effects of freezing the bond length degrees of freedom. Results are discussed which include RMSD plots and statistics. Results are indicating strong correlation between SH and PL.

CHAPTER 7.
EVALUATION/CONCLUSION

We have proposed a so-called penalty function method for constrained molecular dynamics. In this method, a special function is defined so that the function is minimized if the constraints are satisfied. By adding such a function in the potential energy function, the constraints can then be removed from the system, and the simulation can be carried out in a conventional, unconstrained manner. The advantage of using a penalty function method is that it is easy to implement, and does not require solving a nonlinear system of equations in every time step. The disadvantage of the method is that the penalty parameter, i.e., the parameter used to scale the penalty function, is hard to control and in principle, needs to be large enough for the penalty function to be truly effective, and might cause numerical instabilities when used in simulation. It may also arguably be a disadvantage that the penalty function method only force the constraints to be satisfied approximately but not completely. In any case, the method may possibly be used as an alternatively and computationally more efficient approach for constrained molecular dynamics simulation than the Lagrange multiplier types of methods. We first implemented a penalty function method in CHARMM and tested it on protein Bovine Pancreatic Trypsin Inhibitor (BPTI) by following a similar experiment done by Gunsteren and Karplus for the Shake algorithm. In this implementation, we removed the bond length potentials from the potential energy function and introduced the corresponding bond length constraints. For each of the bond length constraints, we constructed a quadratic penalty function and inserted it into the potential energy function. For each different type of bond, we scaled the corresponding penalty function with the force constant of the bond so that the resulting function had the same form as the original bond length potential if without multiplied by the penalty parameter. In this way, the resulting force field becomes simply a continuation of the original force field as the penalty parameter changes continuously from 1 to a value > 1. We conducted a simulation on BPTI with the penalty function method, and compared the results with Verlet and Shake, and found that the penalty function method had a high correlation with the Shake and outperformed the Verlet. In particular, the root-mean-square-deviations (RMSD) of the backbone and non-backbone atoms and the velocity auto correlations of the C_α atoms of the protein calculated by the penalty function method agreed well with those by Shake. The penalty function method requires no more than just applying a conventional, unconstrained simulation algorithm such as the Verlet algorithm to the potential energy function expanded with additional penalty terms for the bond length constraints as stated before. We have also tested the penalty function method on a group of argon clusters with the equilibrium distances for a selected set of molecular pairs as the constraints. Here by the equilibrium distances we mean the distances for the pairs of argon molecules when the clusters are in their global energy minimal states. We generated these distances by using the global energy minimal configuration of the clusters published in previous studies. A penalty function was constructed for each of the constraints and incorporated into the potential energy function of the cluster. The simulation was then conducted by using a conventional, unconstrained simulation method, i.e., the Verlet algorithm, with the extended potential energy function. Here, there were no substantial algorithmic changes or computational overheads required due to the addition of the constraints. The simulation results showed that the penalty function method was able to impose the constraints effectively and the clusters tended to converge to their lowest energy equilibrium states more rapidly than not confined by the constraints. Even if starting structure is out of feasible region, the PL method can be used as a guide to the feasible region and ultimately obtain an optimal solution.

APPENDIX A.
FORTRAN PROGRAM FOR PENALTY FUNCTION METHOD FOR ARGON CLUSTERS

A.0.1 Main program

```
implicit none include "mpif.h"
integer, parameter :: nop=13 ! Number of Particles

real*8, parameter :: velweight=0.5 ! Velocity perturbation

real*8, parameter :: dtt=0.032 ! Time step

real*8, parameter :: time=320.00 ! Total time period

real*8, parameter :: sigma=1 ! STD - to calculate initial velocity

real*8, parameter :: pi=22/7

real*8, parameter :: tol=0.005 ! Tolarence

! Note: if time/dtt divide perfectly is the best choice

real*8, dimension(:),allocatable :: xyz

real*8, dimension(:),allocatable :: xyzout

real*8, dimension(:),allocatable :: velcom

real*8, dimension(:,:),allocatable :: Dis

real*8, dimension(:,:),allocatable :: RDist

real*8, dimension(:,:),allocatable :: Txyz

real*8, dimension(:,:),allocatable :: Txyzout

real*8 :: ttime, T, Pot1, Pot2, lbd, sum

integer :: itno, optat, optat1, optat2

integer :: n, i, j, k, i1, i2, i3, ter
```

```fortran
integer :: m ! number of time steps
integer :: p ! Number of processors
integer :: ierror
integer :: rank ! The rank of the processors
integer :: status(mpi status size)
logical :: minpot
character*4 :: fileno
call MPI INIT(ierror)
call MPI COMM SIZE(mpi comm world, p, ierror)
call MPI COMM RANK(mpi comm world, rank, ierror)
lbd=0 ! Initial penalty parameter value
n=nop
m=int(time/dtt)+1
print*, "Iteration ", m Pot1=100000000.0
allocate(xyz(n*3))
allocate(Txyz(n*3,m))
allocate(xyzout(n*3))
allocate(Txyzout(n*3,m))
allocate(velcom(n*3))
allocate(Dis(n,n))
allocate(RDist(n,n))
itno=0
Txyz=0.0
minpot=.t.
ter=0


CALL read files(nop, Dis, xyz)
```

```
CALL distance(nop, RDist, xyz)
Txyz(:,1)=xyz(:)
CALL Init velocity(dtt, n, sigma, pi, xyz, velcom, T, velweight)
Txyz(:,2)=xyz(:)


do while (minpot)
   Txyzout=0.0
   ttime=mpi wtime()
   open (unit=1,file="Potential.txt",status="new",action="write",iostat=optat1)
   open (unit=3,file="Distance.txt",status="new",action="write",iostat=optat2)
   do k=1,m
      if (k.gt.2) then
        do i=1,n
          i1=3*(i-1)+1
          i2=i1+1
          i3=i1+2
          Txyz(i1,k)=2*Txyz(i1,k-1)-Txyz(i1,k-2)+dtt**2*Txyzout(i1,k-1)
          Txyz(i2,k)=2*Txyz(i2,k-1)-Txyz(i2,k-2)+dtt**2*Txyzout(i2,k-1)
          Txyz(i3,k)=2*Txyz(i3,k-1)-Txyz(i3,k-2)+dtt**2*Txyzout(i3,k-1)
        enddo
      endif
    xyz(:)=Txyz(:,k)
call distance(n, RDist, xyz)
call Verlet(n, xyz, xyzout, dtt, k, sum, Pot2, Dis, lbd)
Txyzout(:,k)=xyzout(:)
if (Pot2.lt.Pot1) then
   Pot1=Pot2
   itno=k
```

```fortran
    endif
    if (mod(k,100)==0) then
      lbd=lbd+0.5 ! increase the Penalty term here
    endif
  enddo
  ttime=mpi wtime()-ttime
  ter=ter+1
  if (sum.lt.tol) then
    minpot=.f.
  endif
  close(3)
  close(1)
enddo
print*, '———— program end normally ————-'
deallocate(Txyz)
deallocate(xyz)
deallocate(xyzout)
deallocate(velcom)
deallocate(Txyzout)
deallocate(Dis)
deallocate(RDist)

call MPI FINALIZE(ierror)

end
```

A.0.2 Sub program Verlet

```
subroutine Verlet(n, xyz, xyzout, dtt, k, sum, Pot, Dis, lbd)
implicit none
real*8, dimension (3*n) :: xyz
real*8, dimension (3*n) :: xyzout
real*8, dimension (n,n) :: Dis
integer :: i, k, j, n, ic, jc, i1, i2, i3, j1, j2, j3, ii, jj
integer :: k, ix, iy, iz, count, casecount
real*8 :: lbd, r1, r2, r3, r, rr, lbda
real*8 :: rv, dtt, Pot, sum, tot, xxyz
real*8 :: xyzsum, xsum, ysum, zsum, r11, r22, r33


xyzout=0.0
Pot=0.0
sum=0.0 count=1
casecount=1
do i=1,n-1
  do j=i+1,n
    lbda=lbd
    i1=3*(i-1)+1
    i2=i1+1
    i3=i1+2
    j1=3*(j-1)+1
    j2=j1+1
    j3=j1+2
    r1=xyz(i1)-xyz(j1)
    r2=xyz(i2)-xyz(j2)
    r3=xyz(i3)-xyz(j3)
    r=(r1**2+r2**2+r3**2)
    xsum=(r-Dis(i,j))*r1
```

```fortran
      ysum=(r-Dis(i,j))*r2
      zsum=(r-Dis(i,j))*r3
      if (Dis(i,j).eq.0.0) then
        lbda=0.0
      endif
      xyzout(i1)=xyzout(i1)+(1/(r**7)-1/(r**4))*r1-lbda*xsum
      xyzout(i2)=xyzout(i2)+(1/(r**7)-1/(r**4))*r2-lbda*ysum
      xyzout(i3)=xyzout(i3)+(1/(r**7)-1/(r**4))*r3-lbda*zsum
      xyzout(j1)=xyzout(j1)-(1/(r**7)-1/(r**4))*r1+lbda*xsum
      xyzout(j2)=xyzout(j2)-(1/(r**7)-1/(r**4))*r2+lbda*ysum
      xyzout(j3)=xyzout(j3)-(1/(r**7)-1/(r**4))*r3+lbda*zsum
      Pot=Pot+(1/(r**6)-2/(r**3))+(1/4)*lbda*(r-Dis(i,j))**2
    enddo
  enddo
  return
  end
```

A.0.3 Sub program Init velocity

```
subroutine Init velocity(dtt, n, sigma, pi, xyz, velcom, T, velweight)
implicit none
real*8, dimension (3*n) :: xyz
real*8, dimension (n) :: vel
real*8, dimension (3*n) :: velcom
real*8, dimension (2*n) :: angle
real*8 :: velweight
real*8 :: sigma
real*8 :: pi
real*8 :: T
real*8 :: dtt
real*8 :: dist
integer :: i
integer :: n
integer :: count
integer :: ccount
T=0.0
count=1
do i=1,3*n,3
   dist=xyz(i)**2+xyz(i+1)**2+xyz(i+2)**2
   vel(count)=(1/sqrt(2*pi*sigma**2)*exp(-dist/(2*sigma**2)))*velweight
   count=count+1
enddo
call random number(angle)
do i=1,2*n,2
   angle(i)=(angle(i)-0.5)*pi
   angle(i+1)=(angle(i+1)-0.5)*pi
enddo count=1 ccount=1
do i=1,3*n,3
```

```
      velcom(i)=vel(count)*sin(angle(ccount))*cos(angle(ccount+1))
      velcom(i+1)=vel(count)*sin(angle(ccount))*sin(angle(ccount+1))
      velcom(i+2)=vel(count)*cos(angle(ccount))
      xyz(i)=xyz(i)+dtt*velcom(i)
      xyz(i+1)=xyz(i+1)+dtt*velcom(i+1)
      xyz(i+2)=xyz(i+2)+dtt*velcom(i+2)
      count=count+1
      ccount=ccount+2
      T=T+(velcom(i)**2+velcom(i+1)**2+velcom(i+2)**2)
   enddo
   T=16*T/n
   return
   end
```

A.0.4 Sub program read files

```
subroutine read files(nop, Dis, xyz)
implicit none

real*8, dimension (nop*3) :: xyz

real*8, dimension (nop,nop) :: Dis

real*8, dimension (nop*3) :: pert

integer :: j, i, k, n

integer :: nop, optat, count

integer :: i1, i2, i3, j1, j2, j3

integer :: ix, iy, iz

integer :: filecount1, filecount2, filecount3

real*8 :: aa, bb, cc

real*8 :: r, r1, r2, r3

real*8 :: xsum, ysum, zsum

character*3 :: fileno

filecount3=0

filecount2=0

filecount1=1

xyz=0

n=nop

Dis=0.0

do k=1,n
   if (mod(k,10)==0) then
     filecount2=filecount2+1
     filecount1=0
   endif
   if (mod(k,100)==0) then
     filecount3=filecount3+1
     filecount2=0
     filecount1=0
```

```fortran
  endif
  filecount1=filecount1+1
enddo
filecount1=filecount1-1
fileno=char(48+filecount3)//char(48+filecount2)//char(48+filecount1)
open (unit=1, file="/usr/people/ajith/LJ/LJ"//fileno//".txt")
j=1
do i=1,n
  read (1,*) (xyz(k),k=j,j+2)
  j=j+3
enddo
close(1)
! Move center to origin xsum=0.0
ysum=0.0
zsum=0.0
do i=1,n-1
  do j=i+1,n
   i1=3*(i-1)+1
   i2=i1+1
   i3=i1+2
   j1=3*(j-1)+1
   j2=j1+1 j3=j1+2
   r1=xyz(i1)-xyz(j1)
   r2=xyz(i2)-xyz(j2)
   r3=xyz(i3)-xyz(j3)
   r=(r1**2+r2**2+r3**2)
   if (i+1==j) then
    Dis(i,j)=r
    Dis(j,i)=r endif
   enddo
   ix=3*(i-1)+1
```

```
  iy=ix+1
  iz=ix+2
  xsum=xsum+xyz(ix)
  ysum=ysum+xyz(iy)
  zsum=zsum+xyz(iz)
enddo
ix=3*(n-1)+1
iy=ix+1
iz=ix+2
xsum=xsum+xyz(ix)
ysum=ysum+xyz(iy)
zsum=zsum+xyz(iz)
xsum=xsum/n
ysum=ysum/n
zsum=zsum/n
do i=1,n
  ix=3*(i-1)+1
  iy=ix+1 iz=ix+2
  xyz(ix)=xyz(ix)-xsum
  xyz(iy)=xyz(iy)-ysum
  xyz(iz)=xyz(iz)-zsum
enddo
! Perturbation
ix=1
call random number(pert)
do i=1,3*n
  if (xyz(i).lt.0) then
    xyz(i)=xyz(i)-pert(ix)*0.3
  elseif (xyz(i).gt.0) then
    xyz(i)=xyz(i)+pert(ix)*0.3
  endif
```

```
  if (mod(i,3)==0) then
    ix=ix+1
  endif
enddo
call random number(pert)
xyz=xyz+(pert-0.5)*.2
return
end
```

A.0.5 Sub program distance

```fortran
subroutine distance(nop, Dist, xyz)
implicit none
real*8, dimension (nop*3) :: xyz
real*8, dimension (nop,nop) :: Dist
integer :: j, i
integer :: nop
integer :: i1, i2, i3, j1, j2, j3
real*8 :: r, r1, r2, r3
do i=1,nop-1
  do j=i+1,nop
   i1=3*(i-1)+1
   i2=i1+1
   i3=i1+2
   j1=3*(j-1)+1
   j2=j1+1
   j3=j1+2
   r1=xyz(i1)-xyz(j1)
   r2=xyz(i2)-xyz(j2)
   r3=xyz(i3)-xyz(j3)
   r=(r1**2+r2**2+r3**2)**(0.5)
   Dist(i,j)=r
   Dist(j,i)=r
  enddo
enddo
return
end
```

APPENDIX B.
HIGH PERFORMANCE FORTRAN PROGRAM FOR PENALTY FUNCTION METHOD FOR ARGON CLUSTERS

```fortran
implicit none
include "mpif.h"

integer, parameter :: nopinx=3 ! Number of Particales in axis (Odd Number)

integer, parameter :: nb=16 ! Number of steps to be done (Book keeping)

 real*8, parameter :: perturb=0.45 ! Initial position pertubation

real*8, parameter :: rm=3.3 ! root mean Error need to maintained

real*8, parameter :: velweight=10.0 ! root mean Error need to maintained

real*8, parameter :: dtxyz=1.7 ! lenght of small cube

real*8, parameter :: dtt=0.032 ! Time step

real*8, parameter :: time=32 ! Total time period

real*8, parameter :: sigma=1 ! STD - to calculate initial velocity

real*8, parameter :: pi=22/7

! Note: if time/dtt divide perfectly is the best choise

real*8, dimension(:),allocatable :: bookkeep

real*8, dimension(:),allocatable :: xyz

real*8, dimension(:),allocatable :: xyzout

real*8, dimension(:),allocatable :: velcom

real*8, dimension(:),allocatable :: Potcollect

real*8, dimension(:),allocatable :: pert

real*8, dimension(:),allocatable :: vel

real*8, dimension(:,:),allocatable :: Txyz

real*8, dimension(:,:),allocatable :: Txyzout

integer, dimension(:),allocatable :: index
```

```fortran
real*8 :: Ta, Tb, velb, vela, ttime, T, Pot1, Pot2, weight, wweight
integer :: decision, itno, isum, optat, n, i, i1, i2, i3, j, k,ter
integer :: m ! number of time steps
integer :: p ! Number of processors
integer :: ierror
integer :: rank ! The rank of the processors
integer :: status(mpi status size)
logical :: minpot
character*4 :: fileno
call MPI/ INIT(ierror)
call MPI/ COMM/ SIZE(mpi comm world, p, ierror)
call MPI COMM RANK(mpi comm world, rank, ierror)
n=(nopinx**2)*nopinx+(nopinx-1)*(nopinx-1)*3*nopinx
m=int(time/dtt)+1
Pot1=100000000.0
isum=0.0
do i=1,n-1
   do j=i+1,n
     isum=isum+i
   enddo
enddo
allocate(bookkeep(isum))
allocate(pert(n*3))
allocate(xyz(n*3))
allocate(vel(n*3))
allocate(Txyz(n*3,m))
allocate(xyzout(n*3))
allocate(Txyzout(n*3,m))
allocate(velcom(n*3))
if (rank==0) then
   allocate(Potcollect(p))
```

```
   allocate(index(p))
endif
decision=99999
itno=0
Txyz=0.0
wweight=perturb/p
do k=1,p
  if (rank==k-1) then
    weight=(k-1)*wweight
  endif
enddo
call Position Init(n, nopinx, dtxyz, xyz, weight)
Txyz(:,1)=xyz(:)
call Init velocity(dtt, n, sigma, pi, xyz, velcom, T, velweight)
Txyz(:,2)=xyz(:)
minpot=.t.
ter=0
do while (minpot)
  bookkeep=1
  Txyzout=0.0
  ttime=mpi wtime()
  do k=1,m
   if (k.gt.2) then
    do i=1,n
      i1=3*(i-1)+1
      i2=i1+1
      i3=i1+2
      Txyz(i1,k)=2*Txyz(i1,k-1)-Txyz(i1,k-2)+dtt**2*Txyzout(i1,k-1)
      Txyz(i2,k)=2*Txyz(i2,k-1)-Txyz(i2,k-2)+dtt**2*Txyzout(i2,k-1)
      Txyz(i3,k)=2*Txyz(i3,k-1)-Txyz(i3,k-2)+dtt**2*Txyzout(i3,k-1)
    enddo
```

```
endif xyz(:)=Txyz(:,k)
call Verlet(n, xyz, xyzout, nb, rm, dtt, k, isum, bookkeep, Pot2)
Txyzout(:,k)=xyzout(:)
if (Pot2.lt.Pot1) then
  Pot1=Pot2
  itno=k
 endif
enddo
ttime=mpi wtime()-ttime
call mpi gather(Pot1,1,mpi real8,Potcollect,1,mpi real8,0,mpi comm world,ierror)
if (rank==0)
 then do i=1,p
   index(i)=i-1
 enddo
 call bubble sort(Potcollect,index, p)
 decision=index(1)
endif
call mpi bcast(decision,1,mpi integer,0,mpi comm world,ierror)
if (rank==decision) then
 Txyz(:,1)=Txyz(:,itno-1)
 Txyz(:,2)=Txyz(:,itno)
endif
if (rank.ne.decision) then
 Txyz=0.0
endif
call mpi bcast(Txyz(:,1),3*n,mpi real8,decision,mpi comm world,ierror)
call mpi bcast(Txyz(:,2),3*n,mpi real8,decision,mpi comm world,ierror)
wweight=1.0/p
do k=1,p
 if (rank==k-1) then
   weight=(k-1)*wweight
```

```fortran
      endif
    enddo
    velb=0.0
    vela=0.0
    do i=1,3*n
      vel(i)=(Txyz(i,2)-Txyz(i,1))/dtt
      if (mod(i,3)==0) then
        velb=velb+(vel(i)**2+vel(i+1)**2+vel(i+2)**2)
      endif
    enddo
    Txyz(:,1)=Txyz(:,2)
    do i=1,3*n
      Txyz(i,2)=Txyz(i,1)+dtt*vel(i)*weight**2
    enddo
    do i=1,3*n
      vel(i)=(Txyz(i,2)-Txyz(i,1))/dtt
      if (mod(i,3)==0) then
        vela=vela+(vel(i)**2+vel(i+1)**2+vel(i+2)**2)
      endif
    enddo
    Tb=(16.0/n)*velb
    Ta=(16.0/n)*vela
    ter=ter+1
    if (ter==10) then
      minpot=.f.
    endif
  enddo
  deallocate(vel)
  deallocate(xyz)
  deallocate(pert)
  deallocate(xyzout)
```

127

```fortran
deallocate(velcom)
deallocate(bookkeep)
if (rank==0) then
   deallocate(Potcollect)
   deallocate(index)
endif
call MPI FINALIZE(ierror)
end
```

B.0.1 Sub program - Verlet

```fortran
subroutine Verlet(n, xyz, xyzout, nb, rm, dtt, k, isum, bookkeep, Pot)
implicit none
real*8, dimension (3*n) :: xyz
real*8, dimension (3*n) :: xyzout
real*8, dimension (isum):: bookkeep
   integer :: k, i, ic, jc, i1, i2, i3, j, j1, j2, j3, n, isum, nb,count, icount
real*8 :: r1, r2, r3, r, rv, rm, dtt, Pot
xyzout=0.0 count=1
if (k.ge.nb) then
  if (mod(k,nb)==0) then
   icount=1
   do ic=1,n-1
    do jc=ic+1,n
     i1=3*(ic-1)+1
     i2=i1+1
     i3=i1+2
     j1=3*(jc-1)+1
     j2=j1+1
     j3=j1+2
     r1=xyz(i1)-xyz(j1)
     r2=xyz(i2)-xyz(j2)
     r3=xyz(i3)-xyz(j3)
     r=(r1**2+r2**2+r3**2)**0.5
     if (rm .gt. r) then
      bookkeep(icount)=1
     else
      bokkeep(icount)=0
     endif
     icount=icount+1
```

```fortran
      enddo
     enddo
    endif
   endif
   Pot=0.0
   do i=1,n-1
    do j=i+1,n
     i1=3*(i-1)+1
     i2=i1+1
     i3=i1+2
     j1=3*(j-1)+1
     j2=j1+1
     j3=j1+2
     r1=xyz(i1)-xyz(j1)
     r2=xyz(i2)-xyz(j2)
     r3=xyz(i3)-xyz(j3)
     r=(r1**2+r2**2+r3**2)**0.5
     if ((bookkeep(count)==1).and.(r.le.2.5)) then
      xyzout(i1)=xyzout(i1)+(1/(r**14)-0.5/(r**8))*r1
      xyzout(i2)=xyzout(i2)+(1/(r**14)-0.5/(r**8))*r2
      xyzout(i3)=xyzout(i3)+(1/(r**14)-0.5/(r**8))*r3
      xyzout(j1)=xyzout(j1)-(1/(r**14)-0.5/(r**8))*r1
      xyzout(j2)=xyzout(j2)-(1/(r**14)-0.5/(r**8))*r2
      xyzout(j3)=xyzout(j3)-(1/(r**14)-0.5/(r**8))*r3
     endif
     count=count+1
     Pot=Pot+(1/(r**12)-2/(r**6))
    enddo
   enddo
   return
   end
```

B.0.2 Sub program - Position Init

```
subroutine Position Init(n, nopinx, dtxyz, xyz, weight)
implicit none

real*8, dimension (3*n) :: seed

real*8, dimension (3*n) :: xyz

real*8, dimension (3*n) :: pert

integer :: rank, p, n, i, ix, iy, iz, j, k, nopinx

real*8 :: dtxyz, cx, cy, cz, x, y, z, weight

z=0.0

xyz=0.0

ix=1

iy=2

iz=3

if (mod(nopinx,2).ne.0) then

   cx=(nopinx-1)/2*dtxyz

   cy=cx

   cz=cx

else

   cx=(nopinx-2)/2*dtxyz

   cy=cx

   cz=cx

endif

do k=1,nopinx

   y=0.0

   do j=1,nopinx

    x=0

    do i=1,nopinx

      xyz(ix)=x

      xyz(iy)=y

      xyz(iz)=z
```

```
        x=x+dtxyz
        ix=ix+3
        iy=iy+3
        iz=iz+3
       enddo
      y=y+dtxyz
     enddo
    z=z+dtxyz
   enddo
   z=0.0
   do k=1,nopinx
    y=dtxyz/2.0
     do j=1,nopinx-1
      x=dtxyz/2.0
      do i=1,nopinx-1
       xyz(ix)=x
       xyz(iy)=y
       xyz(iz)=z
       x=x+dtxyz
       ix=ix+3
       iy=iy+3
       iz=iz+3
       enddo
      y=y+dtxyz
     enddo
    z=z+dtxyz
   enddo
   z=dtxyz/2.0
   do k=1,nopinx-1
    y=dtxyz/2.0
     do j=1,nopinx-1
```

```
       x=0.0
      do i=1,nopinx
        xyz(ix)=x
        xyz(iy)=y
        xyz(iz)=z
        x=x+dtxyz
        ix=ix+3
        iy=iy+3
        iz=iz+3
       enddo
      y=y+dtxyz
     enddo
    z=z+dtxyz
enddo
z=dtxyz/2.0
do k=1,nopinx-1
   y=0.0
   do j=1,nopinx
     x=dtxyz/2.0
     do i=1,nopinx-1
       xyz(ix)=x
       xyz(iy)=y
       xyz(iz)=z
       x=x+dtxyz
       ix=ix+3
       iy=iy+3
       iz=iz+3
      enddo
     y=y+dtxyz
    enddo
   z=z+dtxyz
```

```fortran
enddo
! shifting to all quardrent
do i=1,n
  ix=3*(i-1)+1
  iy=ix+1
  iz=ix+2
  xyz(ix)=xyz(ix)-cx
  xyz(iy)=xyz(iy)-cy
  xyz(iz)=xyz(iz)-cz
enddo
! Perturbation
call random number(pert)
xyz=xyz+
```

B.0.3 Sub program - Init velocity

```fortran
subroutine Init velocity(dtt, n, sigma, pi, xyz, velcom, T, velweight)
implicit none
real*8, dimension (3*n) ::
xyz real*8, dimension (n) :: vel
real*8, dimension (3*n) :: velcom
real*8, dimension (2*n) :: angle
real*8 :: velweight, sigma, pi, T, dtt, dist
integer :: i, n, count, ccount
T=0.0
count=1
do i=1,3*n,3
   dist=xyz(i)**2+xyz(i+1)**2+xyz(i+2)**2
   vel(count)=(1/sqrt(2*pi*sigma**2)*exp(-dist/(2*sigma**2)))*velweight
   count=count+1
enddo
call random number(angle)
do i=1,2*n,2
   angle(i)=(angle(i)-0.5)*pi
   angle(i+1)=(angle(i+1)-0.5)*pi
enddo
count=1
ccount=1
do i=1,3*n,3
   velcom(i)=vel(count)*sin(angle(ccount))*cos(angle(ccount+1))
   velcom(i+1)=vel(count)*sin(angle(ccount))*sin(angle(ccount+1))
   velcom(i+2)=vel(count)*cos(angle(ccount))
   xyz(i)=xyz(i)+dtt*velcom(i)
   xyz(i+1)=xyz(i+1)+dtt*velcom(i+1)
   xyz(i+2)=xyz(i+2)+dtt*velcom(i+2)
```

```
        count=count+1
        ccount=ccount+2
        T=T+(velcom(i)**2+velcom(i+1)**2+velcom(i+2)**2)
    enddo
    T=16*T/n
    return
    end
```

B.0.4 Sub program - bubble sort

```
subroutine bubble sort(A,Index, n)
implicit none

real*8, dimension (n) :: A

integer, dimension (n) :: Index

real*8 :: temp1

integer :: n, i, j, temp2

do j=1,n
  do i=1,n-1
    if (A(i+1).lt.A(i)) then
      temp1=A(i)
      temp2=index(i)
      A(i)=A(i+1)
      index(i)=index(i+1)
      A(i+1)=temp1
      index(i+1)=temp2
    endif
  enddo
enddo
return
end
```

BIBLIOGRAPHY

[1] B. J. Alder and T. E. Wainwright. Phase transition for a hard sphere system. *Journal of Chemical Physics*, 27:2:1208{1209, 1957.

[2] H. C. Andersen. Rattle: A velocity verion of shake algorithm for molecular dynamic calculations. *Journal of computational Physics*, 52:24{34, 1983.

[3] E. Barth, K. Kuczera, B. Leimkuhler, and R. D. Skeel. Algorithms for constrained molecular dynamics. *Journal of Computational Chemistry*, 16:1192{1209, 1995.

[4] E. Barth, B. Leimkuhler, and S. Reich. A test set for molecular dynamics. *Lecture Notes in Computational Science and Engineering*, 24:73{103, 2003.

[5] M. S. Bazaraa, H. D. Sherali, and C. M. Shetty. *Nonlinear Programming; Theory and Algorithms, New York*. John Wiley and Sons, 1993.

[6] D. Beeman. Some multistep methods for use in molecular dynamics calculations. *Journal of computational Physics*, 20:130{139, 1976.

[7] H. M. Berman, J. Westbrook, Z. Feng, G. Gilliland, T. N. Bhat, H. Weissig, I. N.Shindyalov, and P. E. Bourne. The protein data bank. *Nucleic Acids Research*, 28:235{242, 2000.

[8] D. P. Bertsekas. *Constrained optimization and Lagrange multiplirs methods*. Academic Press, New York, 1982.

[9] B. R. Brooks, R. E. Bruccoleri, B. D. Olafson, D. J. States, S. Swaminathan, and M. Karplus. Charmm: A program for macromolecular energy, minmimization, and dynamics calculations. *Journal of computational chemistry*, 4:187{217, 1983.

[10] B. Christianson. Geometric approach to etchers ideal penalty function. *Journal of Optimization Theory and Applications*, 84:433{441, 1995.

[11] R. Courant. Variational methods for the solution of problems of equilibrium and vibrations. *Bulletin of the American Mathematical Society*, 49:1{23, 1943.

[12] L. Dennis. *Advances in Molecular Modeling Volume 1*. Jai press inc, London England, 1990.

[13] L. Dennis. *Advances in Molecular Modeling Volume 2*. Jai press inc, London England, 1994.

[14] L. Dennis. *Advances in Molecular Modeling Volume 3*. Jai press inc, London England, 1995.

[15] G. Di Pillo. *Exact penalty methods in: Algorithms for continuous optimization: the state- of-the-*

art, E. Spedicato (ed.). Kluwer Academic Publishers, Boston, 1994.

[16] J. P. Dussault. Numerical stability and efficiency of penalty algorithms. *SIAM Journal on Numerical Analysis*, 32:296{317, 1995.

[17] R. Fletcher. *A class of methods for nonlinear programming with termination and convergence properties, in: Integer and nonlinear programming, J. Abadie (ed.)*. North-Holland Publishing Company, Amsterdam, London, 1970.

[18] D. B. Gert and V. M. Kurt. *Introduction to molecular dynamics and chemical kinetics*. John Wiley & sons Inc., Canada, 1996.

[19] J. B. Gibson, A. N. Goland, M. Milgram, and G. H. Vineyard. Dynamics of radiation damage. *Physical Review*, 120:1229{1253, 1960.

[20] P. Gierycz and K. Nakanishi. Local composition in binary mixtures of lennard-jones uids with differing sizes of components. *Fluid Phase Equilibria*, 16:255{273, 1984.

[21] P. Gierycz, H. Tanaka, and K. Nakanishi. Molecular dynamics studies of binary mixtures of lennard-jones uids with differing component sizes. *Fluid Phase Equilibria*, 16:241{253, 1984.

[22] P. E. Gill, W. Murray, and M. H. Wright. *Practical Optimization*. Academic press Inc Ltd., London, 1981.

[23] W. F. Gunsteren and M. Karplus. Effect of constraints on dynamics of macromolecules. *Macromolecules*, 15:1528{1544, 1982.

[24] J. P. Hansen and I. R. McDonald. *Theory of simple liquids*. Academic, 1986.

[25] J. P. Hansen and L. Verlet. Phase transitions of the lennard-jones system. *Physical Review*, 184:151{161, 1969.

[26] M. Karplus and J. A. McCammon. Protein structural uctuations during a period of 100 ps. *Nature London*, 277:578, 1979.

[27] J. L. Lagrange. Essai sur une nouvelle methode pour determiner les maxima et minima des formules integrales inde_nies. *Miscellanea Taurinensia II*, pages 173{195, 1762.

[28] A. D. MacKerell, D. Bashford, M. Bellott, R. L. Dunbrack, J. D. Eva seck, M. J. Field, S. Fischer, J. Gao, H. Guo, S. Ha, D. JosephMcCarthy, L. Kuc nir, K. Kuczera, F. T. K. Lau, C. Mattos, S. Michnick, T. Ngo, D. T. Nguyen, B. Pro hom, W. E. Reiher, B. Roux, M. Schlenkrich, J. C. Smith, R. Stote, J. W. Straub, M. tanabe, J. WiorkiewiczKucz- era, D. Yin, and M. Karplus. All-atom empirical potential for molecular modeling and dynamics studies of proteins. *Journal of Physical Chemistry*, B 102:3586{3617, 1998.

[29] A. D. Mackerell, J. Wiorkiewiczkuczera, and M. Karplus. An all-atom empirical energy function for the simulation of nucleic acids. *Journal of the American Chemical society*, Soc. 117:11946{11975, 1995.

[30] J. B. Marion and S. T. Thornton. *Classical dynamics of particles and systems*. Harcourt Brace & Company, 1995.

[31] J. A. McCammon, B. R. Gelin, and M. Karplus. Dynamics of folded proteins. *Nature*, 267:585{590, 1977.

[32] J. A. McCammon, P. G. Wolynes, and M. Karplus. Picosecond dynamics of tyrosine side chains in proteins. *Biochemistry*, 18:927{942, 1979.

[33] J. Meidanis and J. C. Setubal. *Introduction to Computational Molecular Biology*. PWS Publishing Company, 1997.

[34] J. Nocedal and S. J. Wright. *Numerical Optimization*. Springer-Verlag, New York, 1999.

[35] J. A. Northby. Structure and binding of lennard-jones clusters: 13 - 147. *Journal of Chemical Physics*, 87:10:6166{6177, 1987.

[36] J. M. Ortega and W. C. Rheinboldt. *Iterative Solutions of Nonlinear Equations in Several Variables*. Academic Press, 1970.

[37] S. Ozgen and O. Adiguzel. Molecular dynamics simulation of diffusionless phase trans-fomration in a quenched nial alloy model. *Journal of Physics and Chemistry of Solids*, 64:459{464, 2003.

[38] Ryckaert J. P., F. H. Ciccotti, and H. J. C. Berendsen. Numerical-integration of cartesian equations of motion of a system with constraints - molecular-dynamics of n-alkanes. *Journal of computational Physics*, 23:327{341, 1977.

[39] A. Rahman. Correlations in the motion of atoms in liquid argon. *Physics Review*, 136:A:405{411, 1964.

[40] T. Rapcsak. Geodesic convexity in nonlinear programming. *Journal of Optimization Theory and Applications*, 69:169{183, 1991.

[41] T. Rapcsak. *Smooth nonlinear optimization in Rn*. Kluwer Academic Publishers, 1997.

[42] T. Rapcsak. *Global Lagrange multiplier rule and smooth exact penalty functions for equality constraints in: Nonlinear optimization and related topics, eds. G. Di Pillo and F. Giannessi*. Kluwer Academic Publishers, 2000.

[43] D. R. Ronald. A canonocal integration technique. *IEEE transactions on Nuclear Scince*, NS-30 No. 4:2669{2671, 1983.

[44] J.M. Sanz-Serna and Calvo M. P. *Numerical Hamiltonian Problems*. Chapman and Hall, London, 1994.

[45] H. P. Schwefel. *Evolution and Optimum Seeking*. John Wiley and Sons, 1995.

[46] W. C. Swope, H. C. Andersen, P. H. Berens, and K. R. Wilson. A computer simulation method for the calculation of equilibrium constants for the formation of physical clusters of molecules: Application to small water clusters. *Journal of Chemical Physics*, 76:637{649, 1982.

[47] S. Tamar. *Molecular Modeling and Simulation*. New York: Springer Verlag, 2002.

[48] D. J. Tobias and Brooks C. L. Molecular dynamics with internal cordinate constraints. *Journal of Chemical Physics*, 89(8):5115{5125, 1988.

[49] L. Verlet. Computer experiments on classical uids i, thermodynamical properties of lennared jones molecules. *Physical Review*, 159:98{103, 1967.

[50] G. Werner, A. Anton, and C. A. B. Jan. *Fundamental principles of molecular modeling*. New York : Plenum Press, 1996.

[51] L. T. Wille and J. Vennik. Computational complexity of the ground-state determination of atomic clusters. *Journal of Physics*, 18:L:419{422, 1985.

[52] H. Yoshida. Construction of higher order symplectic integrators. *Physics Letters*, A 150:262{268, 1990.

Printed in the United States
By Bookmasters